Sustainable Civil Infrastructures

Editor-in-chief

Hany Farouk Shehata, Cairo, Egypt

Advisory Board

Dar-Hao Chen, Texas, USA
Khalid M. El-Zahaby, Giza, Egypt

About this Series

Sustainable Infrastructure impacts our well-being and day-to-day lives. The infrastructures we are building today will shape our lives tomorrow. The complex and diverse nature of the impacts due to weather extremes on transportation and civil infrastructures can be seen in our roadways, bridges, and buildings. Extreme summer temperatures, droughts, flash floods, and rising numbers of freeze-thaw cycles pose challenges for civil infrastructure and can endanger public safety. We constantly hear how civil infrastructures need constant attention, preservation, and upgrading. Such improvements and developments would obviously benefit from our desired book series that provide sustainable engineering materials and designs. The economic impact is huge and much research has been conducted worldwide. The future holds many opportunities, not only for researchers in a given country, but also for the worldwide field engineers who apply and implement these technologies. We believe that no approach can succeed if it does not unite the efforts of various engineering disciplines from all over the world under one umbrella to offer a beacon of modern solutions to the global infrastructure. Experts from the various engineering disciplines around the globe will participate in this series, including: Geotechnical, Geological, Geoscience, Petroleum, Structural, Transportation, Bridge, Infrastructure, Energy, Architectural, Chemical and Materials, and other related Engineering disciplines.

More information about this series at http://www.springer.com/series/15140

Mounir Bouassida · Mohamed A. Meguid
Editors

Ground Improvement and Earth Structures

Proceedings of the 1st GeoMEast International
Congress and Exhibition, Egypt 2017
on Sustainable Civil Infrastructures

Editors
Mounir Bouassida
Ecole Nationale d'Ingénieurs de Tunis
Université de Tunis EL Manar
Tunis
Tunisia

Mohamed A. Meguid
Faculty of Engineering
McGill University
Montreal, QC
Canada

ISSN 2366-3405　　　　　　　　　ISSN 2366-3413　(electronic)
Sustainable Civil Infrastructures
ISBN 978-3-319-63888-1　　　　　ISBN 978-3-319-63889-8　(eBook)
DOI 10.1007/978-3-319-63889-8

Library of Congress Control Number: 2017946439

© Springer International Publishing AG 2018
This work is subject to copyright. All rights are reserved by the Publisher, whether the whole or part of the material is concerned, specifically the rights of translation, reprinting, reuse of illustrations, recitation, broadcasting, reproduction on microfilms or in any other physical way, and transmission or information storage and retrieval, electronic adaptation, computer software, or by similar or dissimilar methodology now known or hereafter developed.
The use of general descriptive names, registered names, trademarks, service marks, etc. in this publication does not imply, even in the absence of a specific statement, that such names are exempt from the relevant protective laws and regulations and therefore free for general use.
The publisher, the authors and the editors are safe to assume that the advice and information in this book are believed to be true and accurate at the date of publication. Neither the publisher nor the authors or the editors give a warranty, express or implied, with respect to the material contained herein or for any errors or omissions that may have been made. The publisher remains neutral with regard to jurisdictional claims in published maps and institutional affiliations.

Printed on acid-free paper

This Springer imprint is published by Springer Nature
The registered company is Springer International Publishing AG
The registered company address is: Gewerbestrasse 11, 6330 Cham, Switzerland

Preface

Toward building sustainable and longer civil infrastructures, the engineering community around the globe continues undertaking research and development to improve existing design, modeling, and analytical capability. Such initiatives are also the core mission of the Soil-Structure Interaction Group in Egypt (SSIGE) to contribute to the ongoing research toward sustainable infrastructure. This conference series "GeoMEast International Congress and Exhibition" is one of these initiatives.

Ancient peoples built their structures to withstand the test of time. If we think in the same way, our current projects will be a heritage for future generations. In this context, an urgent need has quickly motivated the SSIGE and its friends around the globe to start a new congress series that can bring together researchers and practitioners to pursue "Sustainable Civil Infrastructures." The GeoMEast 2017 is a unique forum in the Middle East and Africa that transfers from the innovation in research into the practical wisdom to serve directly the practitioners of the industry.

More than eight hundred abstracts were received for the first edition of this conference series "GeoMEast 2017" in response to the Call for Papers. The abstracts were reviewed by the Organizing and Scientific Committees. All papers were reviewed following the same procedure and at the same technical standards of practice of the TRB, ASCE, ICE, ISSMGE, IGS, IAEG, DFI, ISAP, ISCP, ITA, ISHMII, PDCA, IUGS, ICC, and other professional organizations who have supported the technical program of the GeoMEast 2017. All papers received a minimum of two full reviews coordinated by various track chairs and supervised by the volumes editors through the Editorial Manager of the SUCI "Sustainable Civil Infrastructure" book series. As a result, 15 volumes have been formed of the final +320 accepted papers. The authors of the accepted papers have addressed all the comments of the reviewers to the satisfaction of the track chairs, the volumes editors, and the proceedings editor. It is hoped that readers of this proceedings of the GeoMEast 2017 will be stimulated and inspired by the wide range of papers written by a distinguished group of national and international authors.

Publication of this quality of technical papers would not have been possible without the dedication and professionalism of the anonymous papers reviewers. The names of these reviewers appear in the acknowledgment that follows. For any additional reviewers whose names were inadvertently missed, we offer our sincere apologies.

We are thankful to Dr. Hany Farouk Shehata, Dr. Nabil Khelifi, Dr. Khalid M. ElZahaby, Dr. Mohamed F. Shehata, and to all the distinguished volumes editors of the proceedings of the GeoMEast 2017. Appreciation is extended to the authors and track chairs for their significant contributions. Thanks are also extended to Springer for their coordination and enthusiastic support to this conference. The editors are thankful for the assistance of Ms. Janet Sterritt-Brunner, Mr. Arulmurugan Venkatasalam in the final production of the 15 edited volumes "Proceedings of GeoMEast 2017".

Contents

**Active Earth Pressure on Retaining Walls
with Unsaturated Soil Backfill** 1
Jagdish Prasad Sahoo and R. Ganesh

**Effect of Geofoam Inclusion on Deformation Behavior
of Buried Pipelines in Cohesive Soils** 20
A.S. Mane, Shubham Shete, and Ankush Bhuse

Response of Strip Footing Adjacent to Nonyielding Basement Wall 34
Magdi El-Emam and Majid Touqan

3D Modeling of EPS Geofoam Buffers Behind Diaphragm Walls 46
Salem A. Azzam, Beshoy M. Shokry, and Sherif S. AbdelSalam

**A Case Study of Efficient Solution for Very High
Geogrid-Reinforced Retaining Wall** 54
Izzaldin M. Almohd, Dimiter Alexiew, and Rami M. El-Sherbiny

**Numerical Simulations of Ground Improvement Using Stone
Columns in "Bouregreg Valley"** 66
Noura Nehab, Khadija Baba, Latifa Ouadif, and Lahcen Bahi

**Behavior of Pipelines Embedded in Self-compacting Materials
Under Traffic Loads** 89
Khalid Abdel-Rahman, Tim Gerlach, and Martin Achmus

**Application of Vacuum Consolidation for the Improvement
of Tunis Soft Soil** 100
Halima Jebali, Wissem Frikha, and Mounir Bouassida

Numerical and Experimental Studies of Sand-Clay Interface 108
Wissem Frikha and Belgacem Jellali

**Effect of the Addition of Chemical Stabilizers
on the Characteristics of Clays** 121
Mouloud Benazzoug and Ramdane Bahar

**Boundary Element Analysis of Shear-Deformable Plates
on Tension-Less Winkler Foundation** 133
Ahmed Fady Farid and Youssef F. Rashed

**Efficient BEM Formulation for Analysis of Plates
on Tensionless Half Space** 146
Marina Reda and Youssef F. Rashed

**Numerical Analysis on the Performance of Fibre Reinforced
Load Transfer Platform and Deep Mixing Columns
Supported Embankments** 157
Liet Chi Dang, Cong Chi Dang, and Hadi Khabbaz

Author Index 171

Active Earth Pressure on Retaining Walls with Unsaturated Soil Backfill

Jagdish Prasad Sahoo[✉] and R. Ganesh

Department of Civil Engineering, Indian Institute of Technology,
Roorkee 247667, India
`jpscivil@gmail.com`, `ravishivaganesh@gmail.com`

Abstract. The active earth pressure coefficients and its distribution against the face of an inclined wall retaining an unsaturated soil backfill, has been established using the limit equilibrium approach. The analysis is performed with the help of a simple Coulomb-type mechanism. For different vertical unsaturated steady state flow conditions, and the location of water table, the variation of soil suction stress that occurs within the vadose zone of the backfill soil mass has been taken into account in the analysis. The influence of different parameters such as inclination of wall, roughness and adhesion of soil-wall interface, ground surcharge pressure, properties of backfill soil and its flow conditions (infiltration, no-flow and evaporation conditions) on the active earth pressure, has been examined in detail. The depth of tensile crack has also been established. The active earth pressure in unsaturated sand is not affected with the variation in the flow conditions for given angle of internal friction of sand; whereas, significant variation in the magnitudes of active earth pressure in unsaturated clay has been observed with the change of rate and type of unsaturated flow. The height of wall and the location of ground water table are found to be the two prime factors that affect substantially the active pressure in unsaturated sand. On the other hand, for a given location of water table the magnitude of the active pressure in unsaturated clay is merely affected by the change in the wall height. The solutions from the present analysis are compared with the available theoretical results that are reported in literature for some special cases.

1 Introduction

Determination of active earth pressure exerted by a backfill soil mass is of considerable importance in the safe design of any retaining structures and helpful partly also for the problems related to sheet piles, braced cuts and foundations etc. Most of the earlier studies deal with this problem assuming the backfill soil mass as either in a state of fully saturated or dry condition. In practice, very often, the backfill soil mass is in unsaturated state. Analysis based on total stress concept for unsaturated soil zone that exist normally above the water table can lead to catastrophic failure of earth structures (Yoo and Jung 2006; Godt et al. 2009; Kim and Borden 2013; Koerner and Koerner 2013). Therefore, the variation of effective stress in the unsaturated soil zone above the water table is required to be established more accurately in order to analyze the stability of

© Springer International Publishing AG 2018
M. Bouassida and M.A. Meguid (eds.), *Ground Improvement and Earth Structures*,
Sustainable Civil Infrastructures, DOI 10.1007/978-3-319-63889-8_1

retaining structures against different possible failures in various stages of its construction. However, the establishment of the variation of effective stress in the unsaturated soil may become difficult owing to the presence of many environmental factors. For instance, flow conditions such as rainfall infiltration or evaporation largely changes the profiles of matric suction and thus, the profiles of effective stress in the unsaturated soil zone are changed. Pufahl et al. (1983) have used two different types of assumptions for defining the nature of matric suction profiles namely, (i) a constant with depth, and (ii) linearly varying with depth in the analysis of earth pressure behind retaining wall with unsaturated backfill. The extended Mohr-Coulomb failure criterion given by Fredlund et al. (1978) was used for describing the maximum shearing resistance offered by unsaturated soil mass; however, this criterion does not consider the nonlinear variation in shear strength due to changes in matric suction beyond air entry value. Also, the effect of change of flux at the ground surface on the shear strength of soil was not taken into account. Nevertheless, the extended Mohr-Coulomb failure criterion given by Fredlund et al. (1978) is still being widely used due to its simplicity for circumstances in which the above mentioned factors are of least important in design. On the other hand, Lu and Likos (2004) have presented modified Mohr-Coulomb failure criterion for describing the shear strength of unsaturated soil; in this failure criterion, both Bishop's effective stress and the suction stress concept were incorporated to define a unified effective stress for saturated as well as unsaturated soil. Therefore, it captures the nonlinear variation in shear strength of unsaturated soils owing to change in matric suction with respect to flow conditions and location of ground water table. As a consequence, this approach enables the realistic nonlinear distribution of active pressure along the length of wall to be predicted. Further, this approach is simple and realistic, satisfying also the flux boundary conditions at the ground surface. The well-known Rankine's earth pressure theory has been extended to unsaturated soils in all the studies mentioned above. Using the definitions of Coulomb earth pressure and applying the unified shear strength theory, Liang et al. (2012) obtained solutions for determining the active earth pressures of unsaturated soils without considering the effect of location of water table and backfill flow conditions. Very recently, Vahedifard et al. (2015) have computed the magnitude of active earth pressure coefficients by employing a log-spiral mechanism with the application of modified Mohr-Coulomb failure criterion.

The above cited studies have clearly demonstrated the importance of explicit consideration of effective stresses in the analysis of retaining walls with unsaturated soil backfill. Although extended Rankine's solution for retaining walls with unsaturated soil backfill has been available for determining active earth pressure distributions, its applicability is only limited to smooth vertical walls. Further, it does not account the effect of ground surcharge pressure. In the present research, by using Coulomb failure mechanism in combination with modified Mohr-Coulomb failure criterion, the active earth pressure coefficients and its distribution against retaining wall with inclined back surface retaining unsaturated soil backfill has been established. The variation of soil suction stress within the vadose zone of the backfill soil mass under various unsaturated flow conditions has been explicitly considered in the analysis. The influence of different unsaturated flow conditions (i.e., infiltration, no-flow and evaporation conditions) and location of water table below base of wall on the magnitude of active earth pressure coefficients and its distribution for varying parameters such as inclination of wall, roughness and adhesion of soil-wall interface, ground surcharge pressure, properties of backfill soil has been examined thoroughly. Finally, the solutions obtained from the present theory have been compared with analytical solutions obtained by Lu and Likos (2004), Vahedifard et al. (2015). It should be noted that (i) extended Rankine's solution presented by Lu and Likos (2004) allows computing only the distribution of active earth pressure against the smooth vertical walls, and (ii) the method suggested by Vahedifard et al. (2015) allows only the magnitude of active earth pressure coefficients to be computed.

2 Problem Statement and Assumptions

A rigid retaining wall of height H with back surface OA is shown in Fig. 1. The back surface of the wall makes an angle of i with the vertical plane. The backfill soil mass is in unsaturated state, and is subjected to a uniform and continuous ground surcharge pressure of magnitude χ. Soil mass is assumed to obey *modified Mohr-Coulomb failure criterion*. The effective values of internal friction angle and cohesion are ϕ' and c', respectively. The water table is assumed to be located at a depth H_0 with respect to the base of wall. During the event of rainfall infiltration and evaporation, vertical one-dimensional unsaturated flow condition is assumed to prevail. Therefore, a constant magnitude of flow rate q at the backfill ground surface is considered with (i) negative sign for steady state infiltration case, and (ii) positive sign for steady state evaporation case. Under no flow condition, the value of q obviously becomes equal to zero. For the

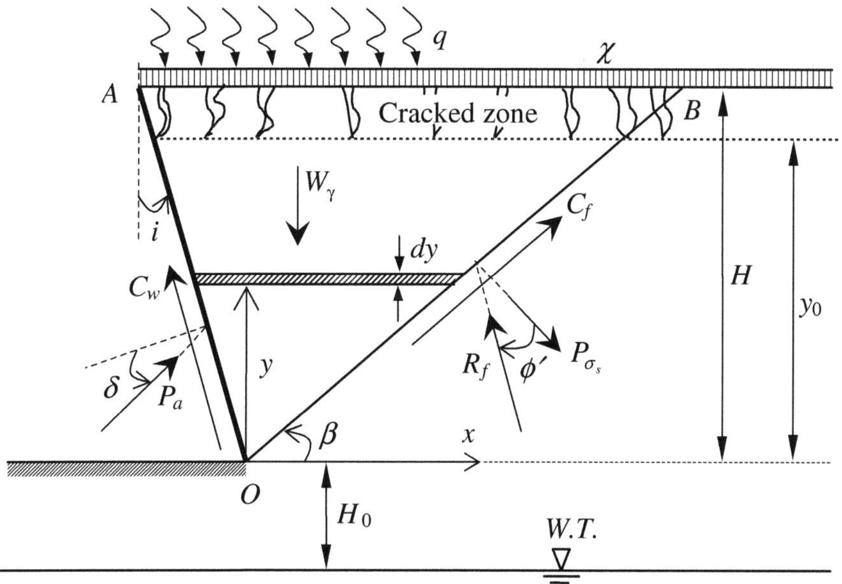

Fig. 1. Details of forces acting on the triangular soil wedge

case of rough walls, the active earth resistance P_a exerted by backfill soil mass is inclined an angle δ with the normal to the wall surface as shown in Fig. 1. A total tangential adhesive force C_w is assumed to be acting at the soil-wall interface due to the wall adhesion component along the length of wall OA. It is aimed to determine the magnitude of total active earth resistance P_a exerted by unsaturated soil backfill on the inclined surface of rigid wall. When the stresses in the backfill soil gets relaxed enough to reach the limiting active state, soil mass is assumed to fail in the shape a triangular wedge OAB with the rupture surface OB making an angle β with the horizontal plane. This simple mechanism is exactly the same as that considered earlier by Coulomb. In order to satisfy the *modified Mohr-Coulomb failure criterion*, the direction of inward reaction force R_f must make an angle ϕ' with the normal to the planar failure surface. The magnitude of total tangential cohesive force C_f acts along the failure surface is simply equal to $c' \times L_{OB}$; where L_{OB} is the length of surface OB. The normal force resulting from suction stress P_{σ_s} acts outwards on the failure surface OB.

3   Method of Analysis

3.1   Shear Strength Based on Modified Mohr-Coulomb Failure Criterion

Following Lu and Likos (2004), on the basis of unified effective stress principle with modified Mohr-Coulomb failure criterion, the shear strength τ_f of saturated or unsaturated soil mass can be written as follows,

$$\tau_f = c' + \sigma' \tan \phi' \tag{1}$$

$$\sigma' = \sigma - u_a + \sigma_s \tag{2}$$

where τ_f defines shear stress at failure, c' and ϕ' are the effective cohesion intercept and effective internal friction angle of soil; σ' represents the effective stress; σ refers to the total stress on the failure plane; u_a is the pore air pressure or is generally considered as atmospheric pressure if not specified which has been taken equal to zero in the present analysis; σ_s represents the suction stress. Considering explicitly the influence of depth of water table H_0 and the properties of soil mass, the suction stress σ_s at any distance measured from the base of the wall in the unsaturated soil backfill under various unsaturated flow conditions can be represented by the following expression (Lu and Likos 2004),

$$\sigma_s = -\frac{1}{\alpha} \frac{\ln\left[\left(1 + \frac{q}{k_s}\right)e^{-\alpha\gamma_w(H_0+y)} - \frac{q}{k_s}\right]}{\left[1 + \left\{-\ln\left[\left(1 + \frac{q}{k_s}\right)e^{-\alpha\gamma_w(H_0+y)} - \frac{q}{k_s}\right]\right\}^n\right]^{\frac{n-1}{n}}} \tag{3}$$

where, k_s is the saturated hydraulic conductivity; γ_w is the unit weight of water; y is the vertical distance from the base of the wall as shown in Fig. 1. α and n are the fitting parameters; α approximates the inverse of the air-entry pressure, and n is directly related to the distribution of the soil's pore size. The possible ranges of α and n for a given soil type can be found in Lu and Likos (2004), Vahedifard et al. (2015). The active earth pressure on retaining walls is affected by the existence of apparent cohesion in the backfill soil mass. The apparent cohesion is defined as the shear strength mobilized by suction stress through the internal angle of friction in the unified effective stress theory, which is expressed as

$$c_a = \sigma_s \tan \phi' \tag{4}$$

where c_a = apparent cohesion; and ϕ' = internal angle of friction.

3.2   Active Earth Pressure in Unsaturated Soil

Under limiting equilibrium condition, by considering horizontal and vertical equilibrium of various forces acting on the failure wedge OAB as illustrated in Fig. 1, the magnitude of active force P_a per unit length of wall, for the case where there is no tensile crack appears in the backfill can be written as,

$$P_a = \frac{(W_\gamma + Q_\chi) \sin (\beta - \phi') - C_w \sin (\beta - \phi' - i) - C_f \cos \phi' - P_{\sigma_s} \sin \phi'}{\cos (\beta - \phi' - \delta - i)} \tag{5}$$

$$C_w = \frac{kc'H \tan \delta}{\cos i \tan \phi'} \tag{6}$$

here, W_γ denotes the weight of soil block OAB and Q_χ is the vertical downward force due to surcharge χ placed on the backfill; k denotes the adhesion factor; $k = 0$ if no wall adhesion is present, i.e., $C_w = 0$ and $k = 1$ when $C_w > 0$. The magnitude of the outward normal force P_{σ_s} due to suction stress can be computed by the following expression,

$$P_{\sigma_s} = \frac{1}{\sin \beta} \int_0^H \sigma_s dy \tag{7}$$

The magnitude of total active earth thrust corresponding to critical failure wedge can be found out by maximizing the function given in Eq. (5) with respect to the angle.

Finally, a unique active earth pressure coefficient in unsaturated soil is defined in the following fashion,

$$K_a = \frac{2}{\gamma H^2} \max(P_a) \big|_{\beta > 0} \tag{8}$$

It is worth mentioning that the magnitudes of neither P_a nor K_a has been corrected for the effect of tension crack zone which would be commonly encountered in the field. The procedure for considering such effects will be discussed in the following section.

3.3 Distribution of Active Earth Pressure

The distribution of earth pressure can be obtained by performing differentiation of the Eq. (5) with respect to the y. In the other words, to obtain the magnitude of active earth pressure p_a exerted by unsaturated soil backfill at any distance y from the base of the retaining wall, the expression for active force P_a given in the Eq. (5) is differentiated with respect to y, which provides,

$$p_a(y) = [\gamma(H-y) + \chi]F_1 - kc'F_2 - c'F_3 - \sigma_s F_4 \tag{9}$$

Where,

$$F_1 = \frac{\cos(\beta - i)\sin(\beta - \phi')}{\cos i \, \sin\beta \, \cos(\beta - \phi'\delta - i)}$$

$$F_2 = \frac{\tan\delta}{\tan\phi'} \frac{\sin(\beta - \phi' - i)}{\cos i \, \cos(\beta - \phi' - \delta - i)}$$

$$F_3 = \frac{\cos\phi'}{\sin\beta \, \cos(\beta - \phi' - \delta - i)}$$

$$F_4 = \frac{\sin\phi'}{\sin\beta \, \cos(\beta - \phi' - \delta - i)}$$

3.4 Depth of Tensile Crack

Because of the negligible or no strength of soils in resisting any tensile stresses, the development of tension crack in the backfill could normally be an unavoidable phenomenon. The maximum possible value of the tensile crack depth $H_t \, (= H - y_0)$ could be developed in the unsaturated soil backfill under active state, can be obtained by satisfying the condition, $y = y_0$ when $p_a(y) = 0$. Therefore, with the usage of Eq. (9) and satisfaction of zero active earth pressure at a distance y_0 from the wall base, the depth of tensile crack H_t can be expressed as follows,

$$H_t = \frac{(kF_2 + F_3)\,c' + \sigma_s\,|_{y=y0}\,F_4}{\gamma F_1} - \frac{\chi}{\gamma} \tag{10}$$

where,

$$\sigma_s\big|_{y=y0} = -\frac{1}{\alpha}\frac{\ln\left[\left(1 + \frac{q}{k_s}\right)e^{-\alpha\gamma_w(H_0 + H - H_t)} - \frac{q}{k_s}\right]}{\left[1 + \left\{-\ln\left[\left(1 + \frac{q}{k_s}\right)e^{-\alpha\gamma_w(H_0 + H - H_t)} - \frac{q}{k_s}\right]\right\}^n\right]^{\frac{n-1}{n}}}$$

The above equation is implicit and hence, the value of H_t from Eq. (10) is obtained herein efficiently by using an iterative process for different parameters such as vertical unsaturated steady state flow rate, depth of water table, properties of soil and wall geometry. The correct solution for H_t is assumed to be obtained when the difference between solutions of two consecutive computations shows negligible value in the order of 10^{-6}.

3.5   Magnitude of Active Earth Pressure with Tensile Crack

Considering the possible occurrence of tensile crack in the backfill under active limiting state, the magnitude of total active resistance P'_a per unit length of wall, can be obtained as follows,

$$P'_a = \int_0^{H-H_t} p_a(y)\,dy \tag{11}$$

On solving and simplifying Eq. (11),

$$P'_a = \frac{H-H_t}{2} \times \left[(2\chi + \gamma(H+H_t))F_1 - 2c'kF_2 - 2c'F_3 - 2P'_{\sigma_s}F_4 \right] \tag{12}$$

where,

$$P'_{\sigma_s} = \frac{1}{H-H_t} \int_0^{H-H_t} \sigma_s\,dy \tag{13}$$

A unique active earth pressure coefficient for the occurrence of tensile crack in the backfill is defined as follows,

$$K'_a = \frac{2P'_a}{\gamma H^2} \tag{14}$$

Thus, the established expressions from the proposed approach allows estimating (i) the magnitude of active earth pressure before and after the formation of tensile crack in the backfill; (ii) the distribution of active earth pressure, and (iii) the maximum possible depth of tensile crack that would occur under active state.

4   Results and Discussion

All the computations required for obtaining the solutions were performed by generating the necessary computer codes. In order to examine the influence of vertical unsaturated steady state flow conditions on the magnitude of active resistance exerted by different types of backfill soils such as sand and clay, the rate of flow q under infiltration, no-flow and evaporation conditions equal to -3.14×10^{-8} m/s, 0 m/s and 1.15×10^{-8} m/s, respectively have been considered on the basis of steady-state infiltration and evaporation rates commonly encountered in the field under natural environmental conditions (Lu and Likos 2004). Different properties of clay and sand used in this analysis are given in Table 1. The typical values of α, n and k_s given in Table 1 are collected from the literature (Lu and Likos 2004; Vahedifard et al. 2015) for the chosen soil types. In order to carry out the parametric study, (i) the cohesion and backfill surcharge pressure have been normalized as $c'/\gamma H$ and $\chi/\gamma H$, respectively, and (ii) the unit weight of chosen soils were kept equal to 20 kN/m^3 irrespective of the soil types. Using Eq. (3) and the properties of soil (Table 1) taken to consideration for performing the analysis, the profile of suction stress has been plotted in Fig. 2(a–b) in terms of normalized suction stress $(\sigma_s/\gamma H)$ vs. normalized distance above water $(y + H_0)/H$ for different wall heights and flow conditions with backfill as clay and sand. It has been found that the suction stress σ_s profiles in unsaturated clay is almost linear and independent of wall height whose magnitude drastically reduces when flow changes from infiltration to evaporation and for all flow conditions, the value of suction stress is always zero at the level of water table and attains its maximum value at the ground surface. The suction stress σ_s profiles in unsaturated sand is highly nonlinear, reduces with a decrease in the height of wall and independent of backfill flow conditions, that is, the stress profiles merge to a single line for all types of flow for a given height of wall. Further, in sand, the value of suction stress increases gradually when the distance from the water table $(y + H_0)$ increases before attaining a peak value after which the value drops continuously and may even get diminishes when the distance $(y + H_0)$ approaches to the ground surface. Similarly using Eqs. (3) and (4), the profile of apparent cohesion for soils under unsaturated steady flow can be obtained, and the profile of both suction stress and apparent cohesion are found to be almost the same with changes only on the magnitudes.

Figures 3 shows the variation of the active earth pressure coefficients, K_a and K_a' with height of wall H under the influence of different backfill flow conditions (infiltration, no-flow and evaporation) for wall inclinations $i = 0°$ with $k = 0$, $H_0 = 0$, $\delta = 1/2\phi'$ and $\chi/\gamma H = 0.1$. It can be seen that the magnitude of K_a and K_a' are found to be continuously increasing with an increase in the height of wall H for unsaturated sand backfill. The values of K_a and K_a' corresponding to different unsaturated flow conditions such as infiltration, no-flow and evaporation has been found to be almost the same for unsaturated sand at all the heights of the wall as a result of which the curves for K_a and K_a' corresponding to different backfill flow in unsaturated sand merged to a single curve. On the other hand, in the case of clay backfill, hardly any change in the values of K_a and K_a' is found to occur with change in the values of H indicating that the pressure

Table 1. Properties of unsaturated clay and sand under consideration

Soil type	ϕ'	$c'/\gamma H$	n	α (kPa^{-1})	k_s (m/s)
Clay	20°	0.05	2	0.005	5×10^{-8}
Sand	30°	0.00	5	0.100	3×10^{-5}

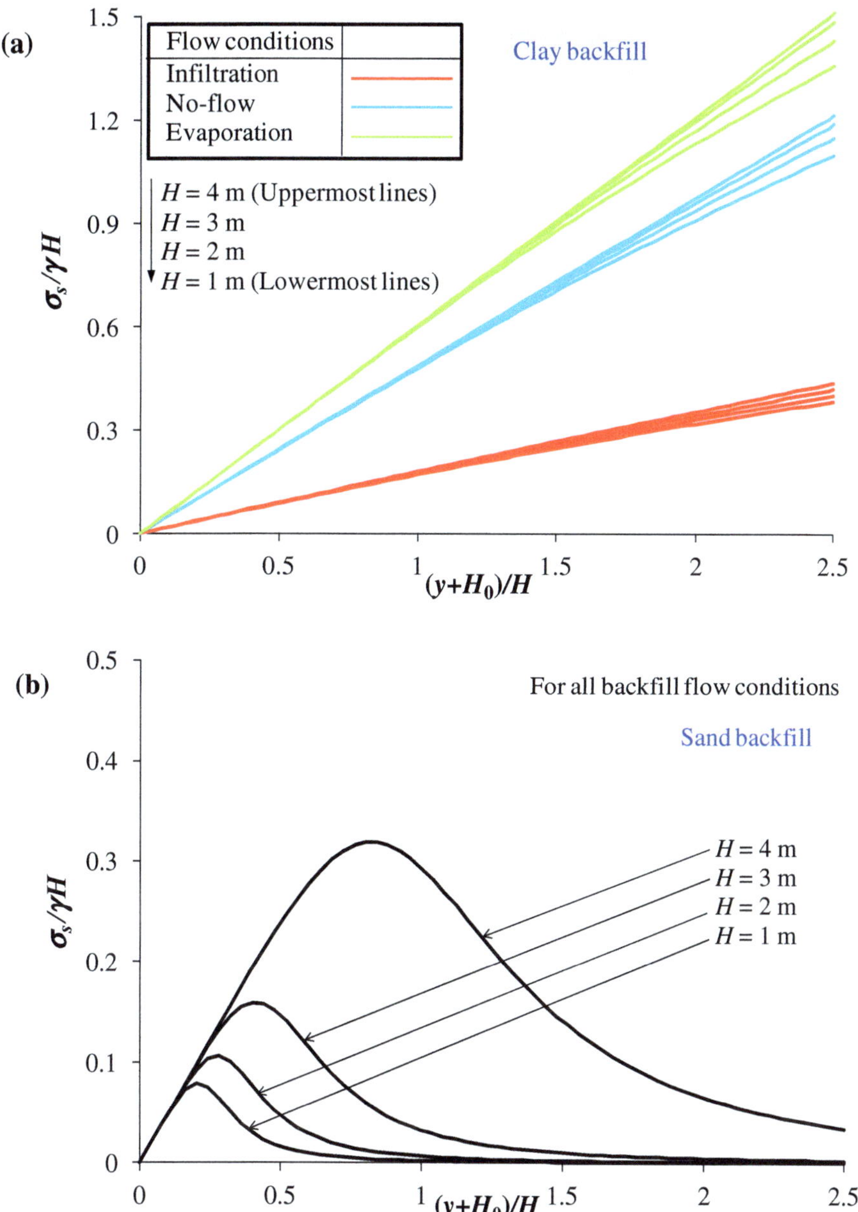

Fig. 2. Variation of normalized suction stress with normalized distance above water for different wall heights and flow conditions with backfill as (a) clay; and (b) sand.

coefficients are independent of the wall height. Moreover, the values of K'_a have been found to be greater than K_a for all the flow conditions and the difference between K'_a and K_a increases when the flow changes from evaporation to no flow to infiltration

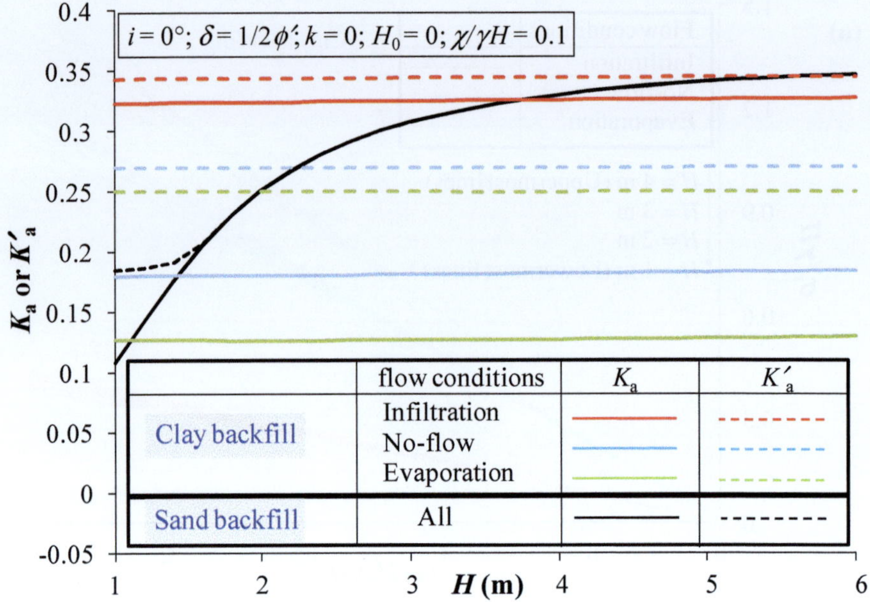

Fig. 3. Variation of active earth pressure coefficients, K_a and K'_a with H for different flow conditions with $i = 0°$, $k = 0$, $H_0 = 0$, $\delta = 1/2\phi'$ and $\chi/\gamma H = 0.1$.

condition. In unsaturated sand, significant difference between the magnitudes of K_a and K'_a is observed for wall height less than 2 m.

For $H_0 = 0$, $\delta = 0$ and $\chi/\gamma H = 0$, the variation of K_a and K'_a in sand backfill with ϕ' (i) for different heights of wall is presented in Figs. 4(a) for $i = -10°$, and (ii) for different inclinations of wall with vertical for H = 3 m is illustrated in Figs. 4(b). The magnitudes of K_a and K'_a decrease continuously with increasing soil friction angle and decreasing height of wall. As aforementioned, it can also be seen from the Fig. 4(a) that the magnitudes of pressure on the retaining wall before and after the crack formation in the backfill mass is exactly the same for wall heights greater than 2 m. For lower heights of wall, the active earth pressure coefficients K_a may even become zero when ϕ' is less than 30°; in such cases the soil backfill has the capability to remain stable without exerting any pressure on the retaining wall. However, the wall may be unstable as active pressure is exerted in the presence of tensile crack. Figures 4(b) shows that the values of K_a and K'_a in sand backfill changes significantly with the inclination of wall. For positive inclination of wall, the magnitude of active pressure on the wall is higher than the negative inclination. This clearly shows that the wall inclination angles have larger impact on the magnitudes of active earth pressure.

The distribution of normalized active earth pressure $p_a/\gamma H$ along the normalized depth of wall y/H for two different surcharges ($\chi/\gamma H = 0$ and 0.1) and for different flow conditions is shown in Figs. 5(a–b) for $H = 3$ m with $k = 0$, $i = 0°$, $H_0 = 0$ and $\delta = 0$. The pressure distribution curves for the case of unsaturated sand are found to be

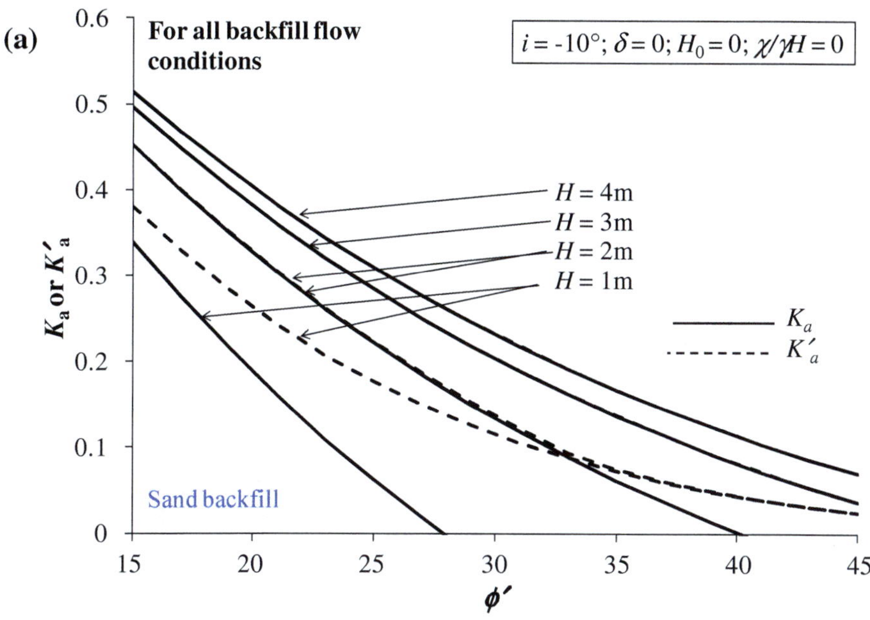

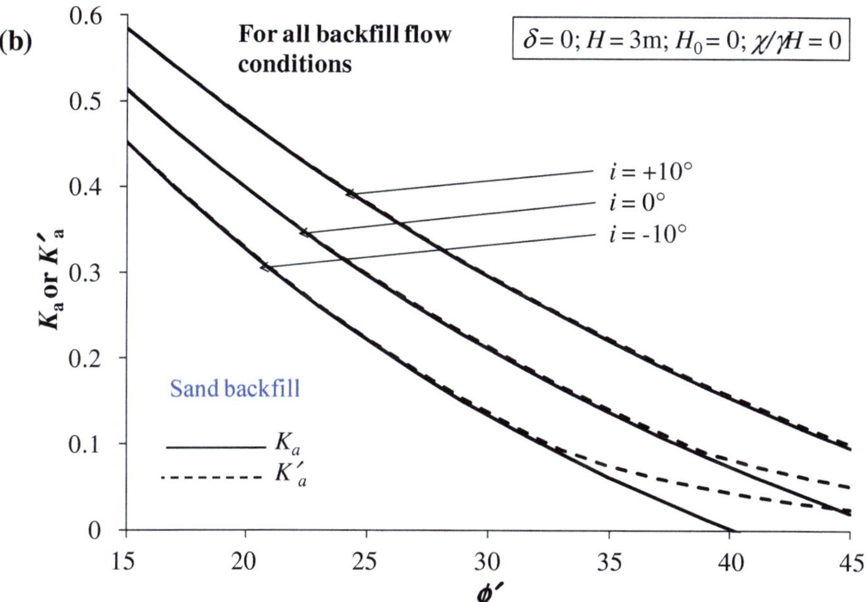

Fig. 4. Variation of active earth pressure coefficients, K_a and K'_a in sand backfill with ϕ' and $H_0 = 0$, $\delta = 0$ and $\chi/\gamma H = 0$ for (a) $i = -10°$ for different flow wall heights; and (b) $H = 3$ m for different vertical inclinations of wall

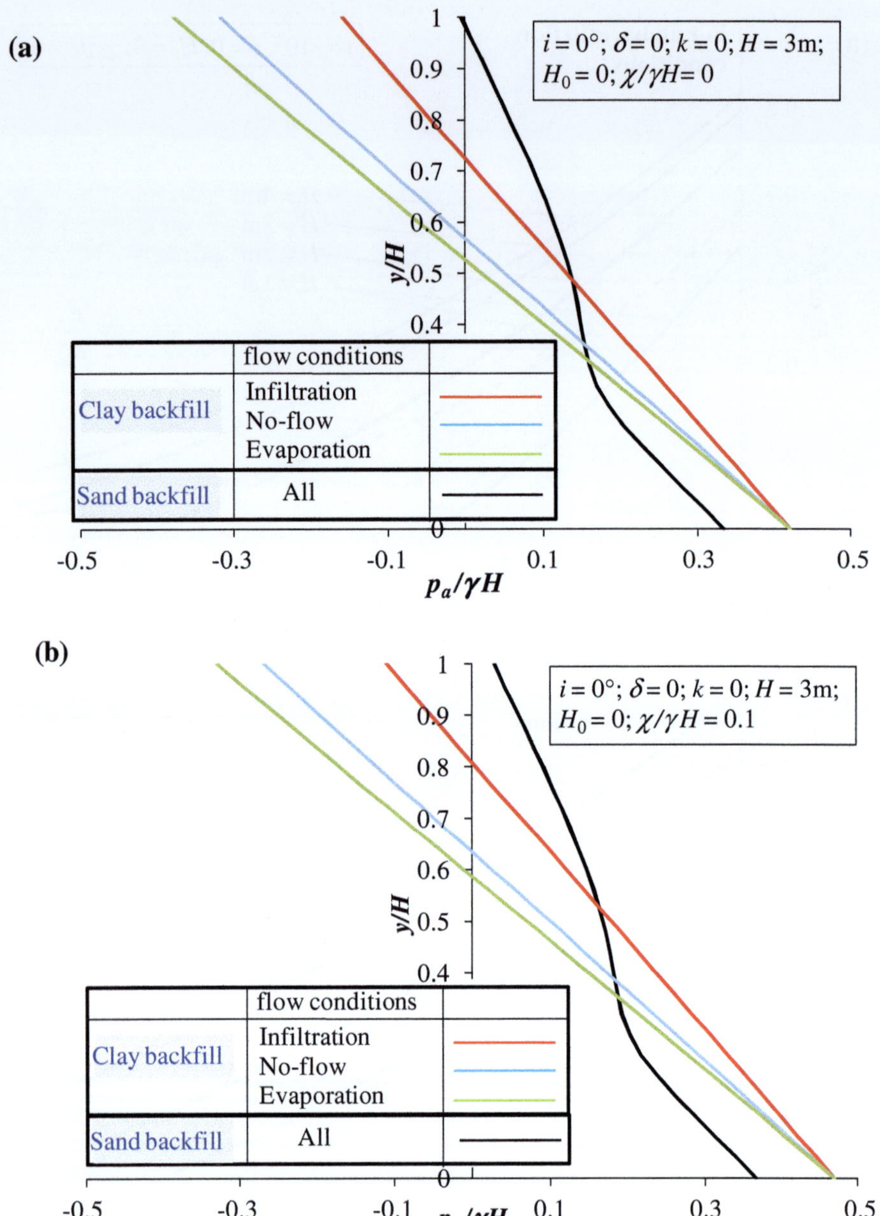

Fig. 5. Distribution of active earth pressure along height of wall for different flow conditions with $k = 0$, $H = 3$ m; $H_0 = 0$, $\delta = 0$ and $i = 0°$ for (a) $\chi/\gamma H = 0$; and (b) $\chi/\gamma H = 0.1$

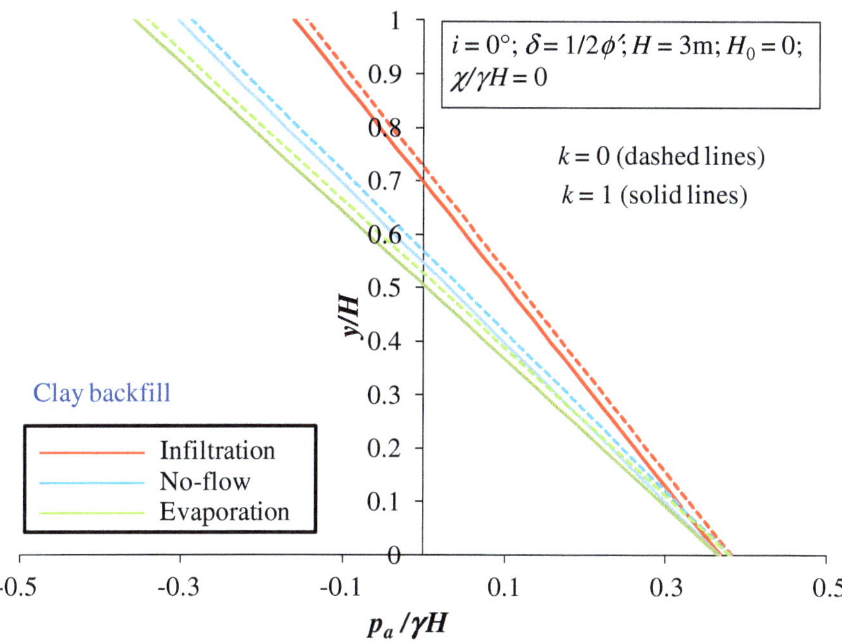

Fig. 6. Distribution of active earth pressure in clay backfill for different flow conditions, and soil-wall adhesion with $\chi/\gamma H = 0$; $H = 3$ m; $H_0 = 0$, $\delta = 1/2\phi'$ and $i = 0°$

nonlinear and not changing with the variation in unsaturated flow conditions. For the case of unsaturated clay backfill, the distribution of $p_a/\gamma H$ along the normalized depth of wall y/H is almost linear in nature. The pressure distribution curves for unsaturated clay backfill tend towards steeper with the flow conditions changing from evaporation to infiltration. It can be noted that the active earth pressure becomes negative in the upper half of the retaining wall for all the flow conditions and the depth of tensile crack zone (equal to the depth at which the pressure curves changes its sign) is found to increase when the backfill flow changes from infiltration to evaporation. Therefore, the magnitudes of K'_a are noticed to be maximum and minimum for the case of infiltration and evaporation, respectively. The increase in the magnitude of backfill surcharge pressure causes the pressure distribution curves to shift towards the positive axis resulting an increase in the values of K_a and K'_a.

For studying the effect of soil-wall interface characteristics on the active earth pressure distribution, Figs. 6 and 7 are plotted for $\chi/\gamma H = 0$, $H = 3$ m, $i = 0°$ and $H_0 = 0$ corresponding to different backfill flow conditions. It can be seen from the Fig. 6 that the increase in the wall adhesion from its possible minimum ($k = 0$) to maximum ($k = 1$) values tend to shift the pressure distribution curves slightly towards the negative pressure axis, which indicates a marginal decrease in the pressure coefficients corresponding to different flow conditions. The increase in the soil-wall interface friction angle (δ) reduces the magnitude of active earth pressure. It is worth mentioning that the change of flow conditions and soil-wall interface friction angle

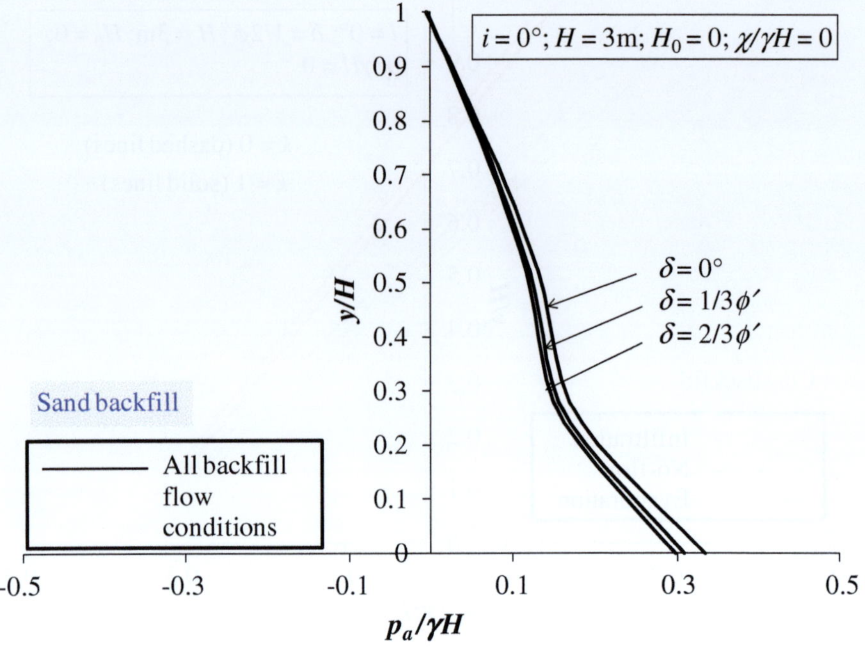

Fig. 7. Distribution of active earth pressure in sand backfill along height of wall for different flow conditions, and soil-wall interface friction angles with $\chi/\gamma H = 0$; $H = 3$ m; $H_0 = 0$ and $i = 0°$

does not seem to alter the pressure distribution curves near the top of wall in the unsaturated sand backfill.

In order to examine the effect of location of water table (H_0) on the active pressure distribution, for different normalized values of location of water table H_0/H by taking the values of parameters $\chi/\gamma H = 0$, $H = 3$ m, $\delta = 0$ and $i = 0°$, the variation of $p_a/\gamma H$ along the normalized depth of wall y/H is shown in Figs. 8 and 9 for unsaturated sand and clay as backfill, respectively. For a given orientation, heights of wall and backfill properties of sand, the pressure distribution curves in the sand backfill (Fig. 8) are highly non-linear for lower values of $H_0/H < 1/3$, and nearly linearly active earth pressure profiles were observed for larger values of H_0/H. As aforementioned and referring to Fig. 2, the suction stress reaches peak value at a relatively lower height above the water table which in turn causes the active earth pressure profile to attain a nearly linear trend either with increasing the height of wall or lowering of water table. The magnitudes of active earth pressure gradually increases with an increase in the values of H_0/H. Thus, in practical situations, it may be expected that the height and location of water table has significant effect either on reducing or on increasing the values of active earth pressure depending the geometry of wall and properties of sand backfill. Figures 9 shows in the case of clay backfill, the pressure distribution curves are shifted towards the negative pressure axis with increasing in H_0/H for all flow conditions. Unlike sand backfill, a greater dependency of flow type on changing the

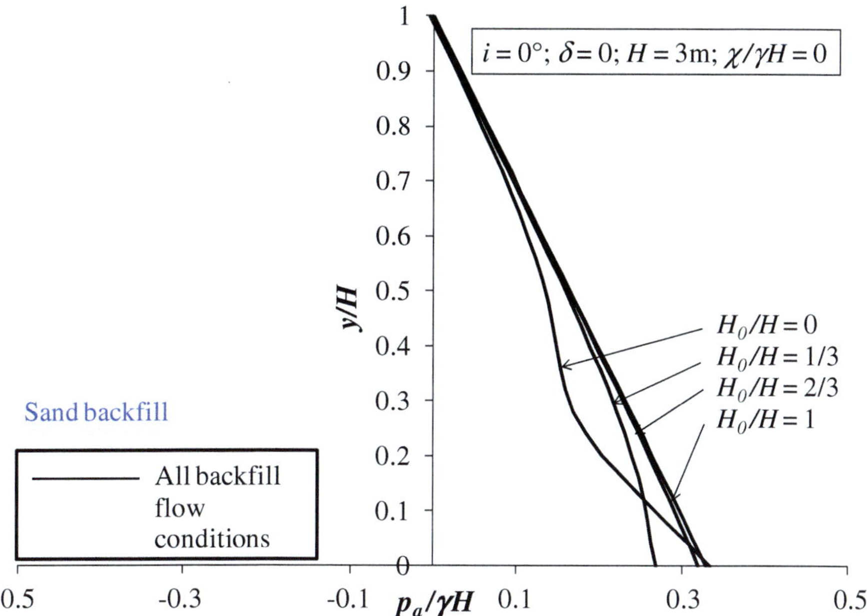

Fig. 8. Distribution of active earth pressure in sand backfill along height of wall for different flow conditions, and water table positions from the wall base with $\chi/\gamma H = 0$; $H = 3$ m; $\delta = 0$ and $i = 0°$

profiles of active earth pressure in clay backfill is observed and the reduction in the magnitude of earth pressure is found to increase with changing flow condition from infiltration to evaporation. As the variation of suction stress profiles between the water table and the top surface of retained soil mass at a given flow rate is almost linear in unsaturated clay, any increase in the value of H_0/H will obviously lead to reduction in the magnitude of earth pressure.

5   Comparison with Available Solutions

The active earth pressure coefficients obtained from the present analysis has been compared with the solutions reported by Vahedifard et al. (2015) for an inclined wall ($i = -10°$) with $H = 4$ m, $H_0 = 0$, $\chi/\gamma H = 0$, $\delta = 0$ for two different backfill soils. The associated comparisons are presented in Fig. 10(a–b). It can be noticed from Fig. 10(a) that the magnitudes of active earth pressure coefficients obtained (K_a) for unsaturated sand backfill are slightly higher (critical) than that reported by Vahedifard et al. (2015) using log spiral slip surface. It is noteworthy that no tensile crack could be developed in the backfill soil mass under limiting active state based on Eq. (10). For clay backfill as shown in Fig. 10(b), the present solutions for active earth pressure coefficients K_a and

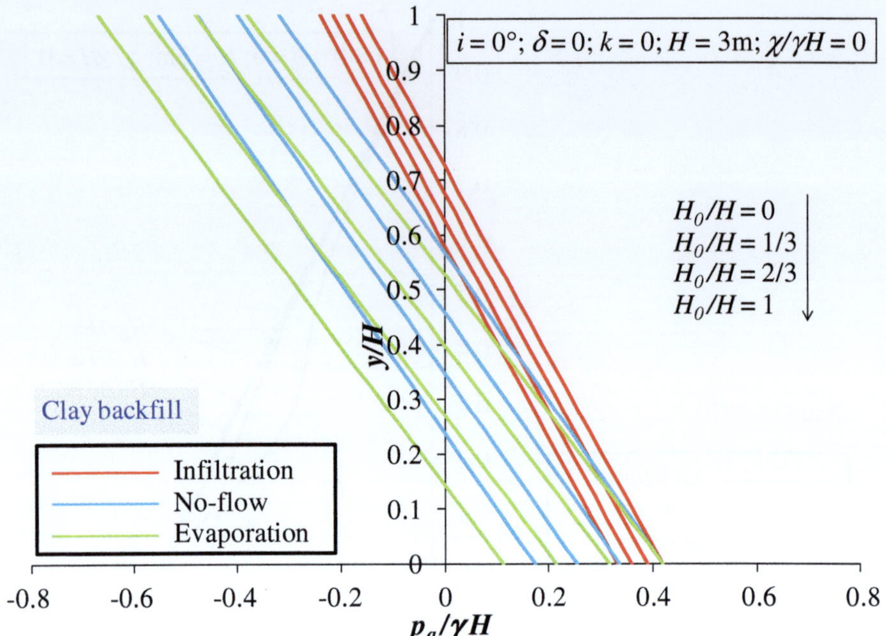

Fig. 9. Distribution of active earth pressure in clay backfill along height of wall for different flow conditions, and water table positions from the wall base with $\chi/\gamma H = 0$; $H = 3$ m; $\delta = 0$; $k = 0$ and $i = 0°$

K'_a obtained with and without considering the effects of tensile crack has been compared with K_a values reported by Vahedifard et al. (2015). It is found that the values of K_a provided by Vahedifard et al. (2015) lies in between the K_a and K'_a values obtained in the present analysis. The difference in two solutions may be expected due to the assumption followed by Vahedifard et al. (2015) in defining the point of application of active pressure.

In Fig. 11(a–b), the active earth pressure profiles established from the present analysis are compared with solutions of Lu and Likos (2004) for two different unsaturated backfill soils retaining by a smooth vertical wall with $H_0 = 10$ m, $H_0 = 0$, $\chi/\gamma H = 0$, $k = 0$ and $\delta = 0$. It can be clearly seen that the present solution are exactly the same to that reported by Lu and Likos (2004) based on extended Rankine's theory. In case of a smooth vertical wall with $k = 0$, the value of critical failure plane corresponding to maximum active pressure is found to be $45 + \phi'/2$. Thus, the solutions from the present method for a vertical smooth wall with $k = 0$ and the extended solutions provided by Lu and Likos (2004) remains exactly the same.

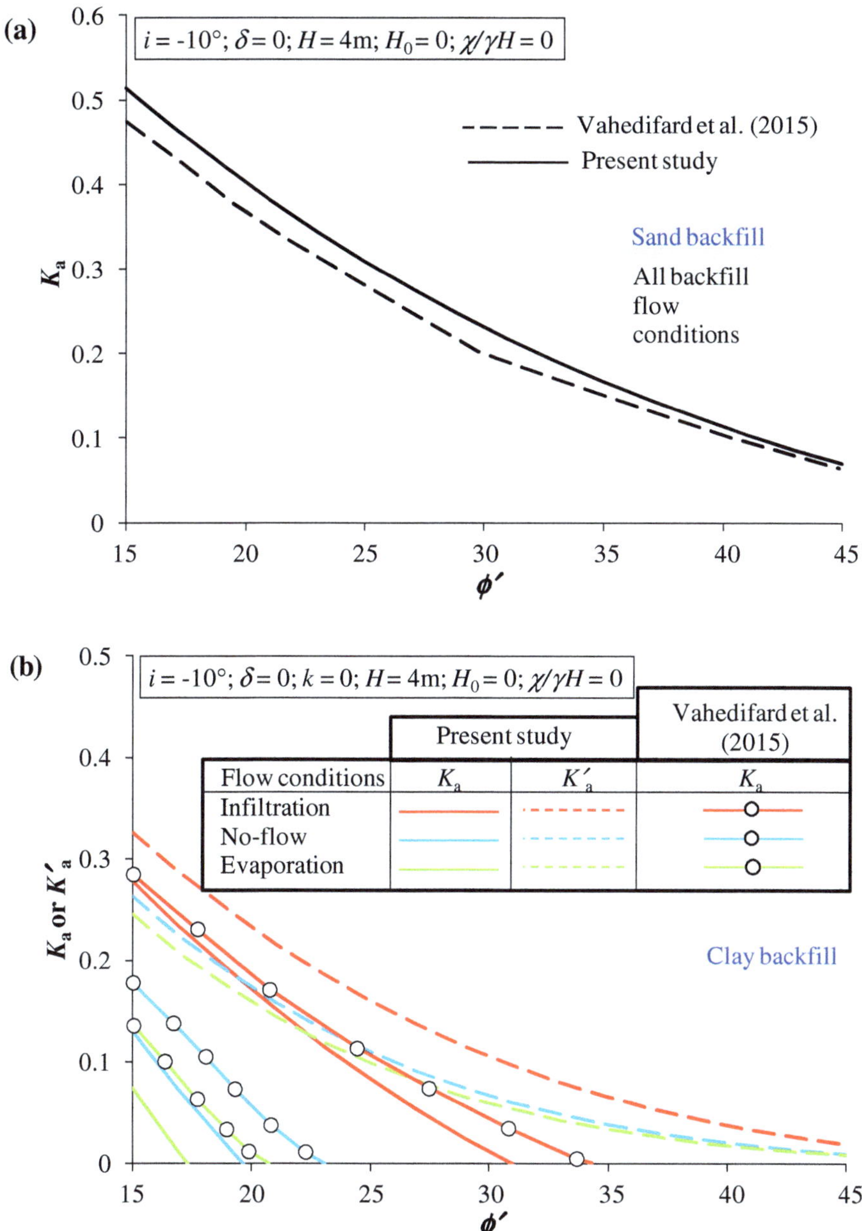

Fig. 10. Comparison of active earth pressure coefficients from the present study with that reported by Vahedifard et al. (2015) for (a) sand backfill; (b) clay backfill

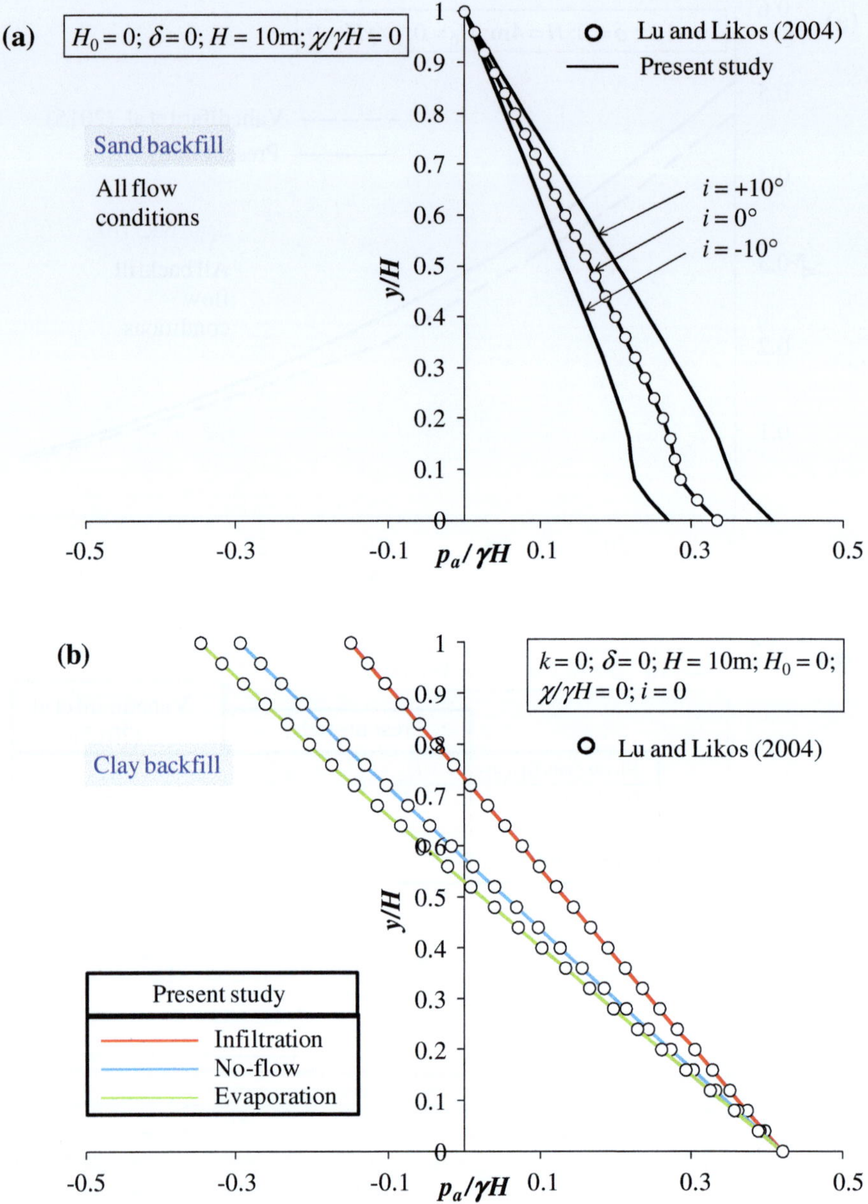

Fig. 11. Comparison of active earth pressure distribution against vertical smooth walls for (a) sand backfill, and (b) clay backfill under different backfill flow conditions from the present study with that reported by Lu and Likos (2004)

6　Conclusions

With the consideration of simple Coulomb failure mechanism based on modified Mohr-Coulomb yield criterion, the active earth pressure coefficient and its distribution against retaining walls with unsaturated backfill soils has been presented in the framework of limit equilibrium analysis. The effect of different vertical flow conditions and the location of water table with respect to base of wall on the active earth pressure exerted by unsaturated clay and sand backfill have been studied by varying different parameters such as inclination of wall, roughness and adhesion of soil-wall interface. The present approach allows computing the active pressure developed due to unsaturated backfill before and after the occurrence of tensile crack. For a given internal friction angle of soil, wall orientation and soil-wall interface property, the active earth pressure in unsaturated sand is found to be largely dependent on the height of wall and also the location of water table; whereas, in unsaturated clay, it is dependent on the location of ground water table. The variation in flow conditions alters more significantly the magnitude of resultant active earth thrust against the wall with unsaturated clay backfill; however, for unsaturated sand backfill, the magnitude of resultant active force remains unaffected with changes in the backfill flow conditions. The present study could be helpful to understand the causes of failure of retaining structures under different flow conditions in the presence of unsaturated backfill.

References

Fredlund, D.G., Morgenstern, N.R., Widger, R.A.: The shear strength of unsaturated soils. Can. Geotech. J. **15**, 313–321 (1978)

Godt, J.W., Baum, R., Lu, N.: Landsliding in partially saturated materials. Geophys. Res. Lett. **36** (2), L02403 (2009)

Kim, S.K., Borden, R.H.: Numerical simulation of MSE wall behavior induced by surface-water infiltration. J. Geotech. Geoenviron. Eng. **139**(12), 2110–2124 (2013)

Koerner, R.M., Koerner, G.R.: A data base, statistics and recommendations regarding 171 failed geosynthetic reinforced mechanically stabilized earth (MSE) walls. Geotext. Geomembr. **40**, 20–27 (2013)

Liang, W., Zhao, J., Li, Y., Zhang, C., Wang, S.: Unified solution of Coulomb's active earth pressure for unsaturated soils without crack. Appl. Mech. Mater. **170–173**, 755–761 (2012)

Lu, N., Likos, W.J.: Unsaturated Soil Mechanics. Wiley, New York (2004)

Pufahl, D., Fredlund, D., Rahardjo, H.: Lateral earth pressures in expansive clay soils. Can. Geotech. J. **20**(2), 228–241 (1983)

Vahedifard, F., Leshchinsky, B.A., Mortezaei, K., Lu, N.: Active earth pressures for unsaturated retaining structures. J. Geotech. Geoenviron. Eng. **141**(11), 04015048 (2015)

Yoo, C., Jung, H.: Case history of geosynthetic reinforced segmental retaining wall failure. J. Geotech. Geoenviron. Eng. **132**(12), 1538–1548 (2006)

Effect of Geofoam Inclusion on Deformation Behavior of Buried Pipelines in Cohesive Soils

A.S. Mane[1](✉), Shubham Shete[2], and Ankush Bhuse[2]

[1] Department of Civil Engineering, CSMSS,
Aurangabad 431005, Maharashtra, India
abhinavmane@gmail.com
[2] DIEMS, Aurangabad 431005, Maharashtra, India
shubhamshete1231@gmail.com, ankushbhuse94@gmail.com

Abstract. The response of buried pipelines in cohesive soils with and without geofoam inclusion was studied extensively in this paper. Evaluation was made with the help of small-scale model tests. A series of small-scale models was performed in a fabricated box test setup, which defines the buried pipeline in cohesive soil. Black cotton soil emerged from basaltic formation in Maharashtra region of India was chosen to represent cohesive fill over and around buried pipes maintaining the constant embedment depth. Fabricated test setup was equipped with the front transparent glass panel to facilitate the capture of particle movements in the small-scale model during the increments of the loading. A 2-inch diameter HDPE pipe was used so as to represent prototype buried pipes. Geofoam was used as a compressible inclusion varying its density and cross sectional width. Plane strain conditions were adopted for all the tests. An image analysis technique was used to evaluate the performance of the geofoam in enhancement of deformation behavior of the buried pipe. Strip loading was applied with a constant load rate of 0.1 N/Sec using a Universal Testing Machine (UTM). This facilitates the correct evaluation of dissipation of the energy due to geofoam through soil arching and compression of the geofoam. Inclusion of geofoam around buried pipe prevents the adverse effects of unforeseen excessive forces on the pipeline resulting in minimal serviceability of the pipelines, reduced cost of maintenance, and reduced losses in the system and finally the effective economical operations in adverse geotechnical conditions. A maximum reduction of 32.14% was observed in the vertical deformations of buried pipe when a 150 mm wide low density geofoam was included beneath the shallow foundation at embedment depth equal to width of the footing.

Keywords: Buried pipelines · Geofoam · Black cotton soil · Small scale modeling · Image analysis

1 Introduction

The pipelines usually buried below the ground for economic, aesthetic and environmental reasons. Generally oil and gas pipelines are designed and constructed as continuous pipelines, while water supply pipelines are constructed as segmented pipelines.

© Springer International Publishing AG 2018
M. Bouassida and M.A. Meguid (eds.), *Ground Improvement and Earth Structures*,
Sustainable Civil Infrastructures, DOI 10.1007/978-3-319-63889-8_2

These pipelines are subjected to different types of loads like soil load, traffic load and construction loading also. Due to the temperature and pressure pipelines also change its behavior as well as different types of soil would also affect on pipeline. Several authors have put forward their studies Corey et al. (2014); Stephen (2011); (Watkins (2004); Johnson et al. (2010); Anirban De and Zimmie (2016); Lin and Chou (2012)). Behavior of fine grained cohesive soils varies greatly when it comes contact with the water. It is absolutely essential to shield from destruction the buried pipelines against differential action of loading. The different types of load may cause to unreasonable differential deformation in pipes, which may further result in damage or crack of such pipes, disruption in the transportation or intended fluids. Some of the authors have suggested the use of geofoam to protect these pipes from surrounding soil (Bilgin and Stewart (2012)). However, their research is mainly limited to the concept-based application than the actual modeling of the buried pipelines. Buried pipelines could be protected by using geofoam inclusion with varying effectiveness of several factors such as density and width of the geofoam, etc. Present study demonstrates the small-scale experimental evaluation of the buried pipelines with and without geofoam. Three different densities of geofoam was adopted in the present study along with the three different widths. The model pipe diameter and loading type were kept constant throughout the study. A strip load 0.203H wide (where, H is the height of the pipe embedment) was applied on each of the model test at the top surface of the soil. The width of the strip load was chosen such that the Terzaghi's failure does not extend upto the pipe embedment depth.

2 Motivation Behind Present Study

Black cotton soil is highly unreliable in nature due to presence of clay minerals especially montmorillonite. When strip load is gradually applied on soft clay, it forms a zone of punching shear failure. The punching shear failure directly transfers the load on buried pipe causes deformation in the pipe. Figure 1 shows the schematic cross section of the buried pipes with and without geofoam inclusion. The zone of punching shear extends the deformation of clay and load deposited towards the pipe. When geofoam placed below the strip footing at a certain depth, geofoam compresses and the settlement due to strip loading get distributed in the surrounding clay. This forms an uniform arching in clay which distributes the deformations in the fine grained soil above the pipe. Which further provides higher load bearing area and thus comparatively less load transferred on the buried pipe. Spreading in load transfer should heighten with compressibility of the geofoam as well as with the available volume for compression under the foundation. Higher compressibility could be achieved with decrease in geofoam density, providing higher volume change in the embedment area. Increase in width of geofoam may demonstrate possible efficacious results in load scattering. With higher width of geofoam below footing wider the spread of load and greater possibilities for clay arching, increase bearing area and shear strength development of clay. So, the present study demonstrates the evaluation of effectiveness of geofoam inclusion to reduce earth pressure on buried pipes in cohesive soil. Small-scaled experimental evaluation was performed with parametric variation in geofoam width (50, 100 and 150 mm) and geofoam density (8, 16 and 24 kg/m^3).

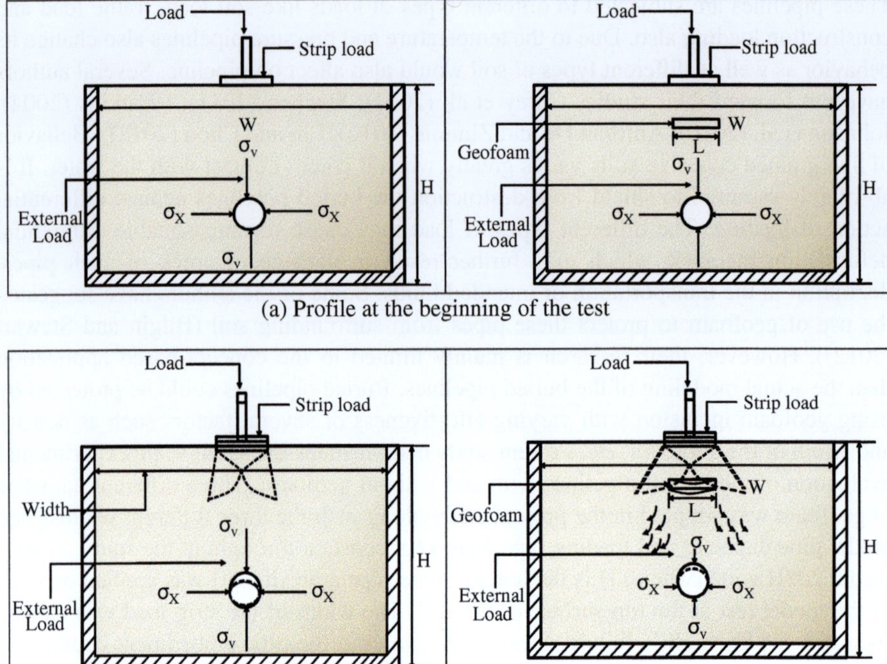

Fig. 1. Schematic cross section of buried pipe with and without geofoam

3   Model Materials

3.1   *Soil*

Black cotton soils are found in many part of India. The black cotton used in this study was collected from Mukundwadi, Aurangabad (Maharashtra). Black cotton soils have been identified on igneous, sedimentary and metamorphic rocks. It is mainly formed by the chemical weathering of igneous rock namely basalt. The soil is classified as high plasticity clay (CH) with expansive behavior. The optimum moisture content of black cotton was found 18% at maximum dry density of 1.56 g/cm^3. In the present study, soil was placed at 10% wet of optimum conditions. This was obtained as dry density 1.404 g/cm^3 and moisture content 29%. Summary of properties of the model black cotton is shown in Table 1. Figure 2 (a) shows the pictorial view of collected sample of black cotton soil.

3.2   *Geofoam*

To represent a compressible inclusion below the shallow strip footing, expanded polystyrene (EPS) geofoam was used. EPS8, EPS16 and EPS24 types of geofoam were used in the present study with varying density 8, 16 and 24 kg/m^3 respectively.

Table 1. Properties of model materials used in present study

Particulars	Quantity		
Black cotton soil			
Soil classification	CH		
Liquid limit (%)	68		
Plastic limit (%)	26.40		
Plasticity index (%)	43.50		
Optimum moisture content (%)	18		
Maximum dry density (gr/cc)	1.56		
Grain size distribution: Sand (%)	5.1		
Silt (%)	45		
Clay (%)	49.9		
HDPE pipe			
Diameter, D_p(m)	0.05		
Compressive load at 10% strain, (kN)	26		
Geofoam			
Geofoam type	Expanded polystyrene	Expanded polystyrene	Expanded polystyrene
Geofoam legend	EPS8	EPS16	EPS24
Density (kg/m^3)	8	16	24
Compressive resistance at 2% strain (kPa)	17	42	74
Compressive elastic modulus (kN/m^2)	850	2100	3700

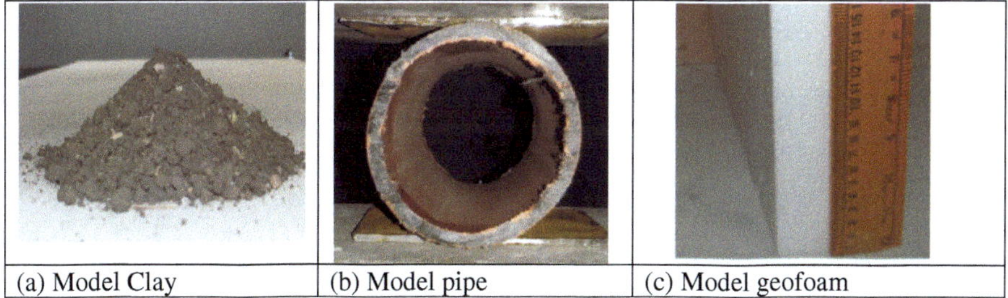

| (a) Model Clay | (b) Model pipe | (c) Model geofoam |

Fig. 2. Photograph view of model materials used in the present study

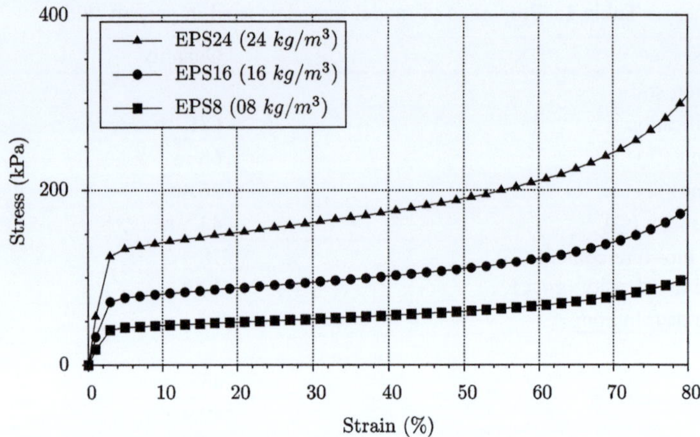

Fig. 3. Unconfined uniaxial stress strain behavior of model geofoam

The unconfined compressive resistance and elastic modulus was obtained as to be 17, 42, 74 kPa and 850, 2100 and 3700 kPa for geofoam EPS8, EPS16 and EPS24 respectively. The unconfined compressive stress strain variation for model geofoam used in the present study shown Fig. 3. Table 1 summarizes properties of the model geofoam.

3.3 HDPE pipe

The model pipe stands for the flexible buried pipeline used for water, gas and oil transportation. In the present study a commercially purchasable 2-inch diameter flexible HDPE pipe was used. Properties of the model pipe were assessed through uniaxial compression test. A simple arc arrangement was made with mild steel so as to ensure the load transfer in single vertical axis and no slippage occurs during the compression

(a) At beginning of the test

(b) At the end of the test

Fig. 4. Photograph view of uniaxial compression test on model HDPE pipe

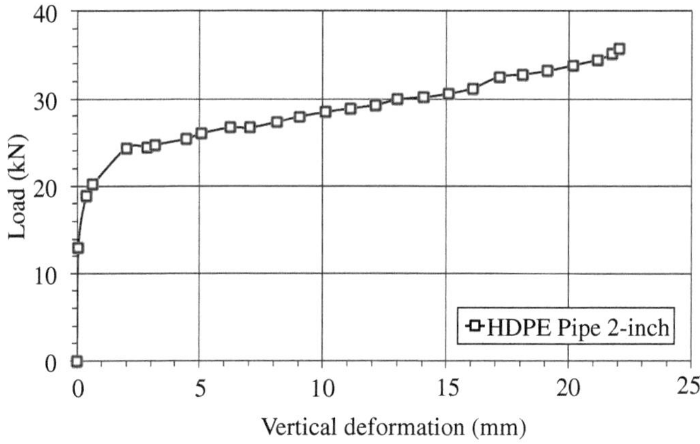

Fig. 5. Uniaxial load deformation variation of model HDPE pipe

tests. Photographic view of uniaxial compression test on model HDPE pipe used in the present study is shown in Fig. 4. Variation of load with vertical deformation of the pipe is shown in Fig. 5. Correlation of deformation to the vertical load was further used in the analysis and rendering to assess the load transferred on the pipe with and without geofoam inclusion. The loads were estimated on buried pipes based on the uniaxial load deformation information and are representative in nature. The approach followed was to evaluate the efficacy of the geofoam and limited to uniaxial tests in the present study. However, to arrive upto the accurate load transferred over the buried pipes the confined load deformation relation should be adopted, which is the future scope for the present study.

4   Model Test Package and Test Procedure

4.1   *Model Test Package*

The front view of the buried pipeline model constructed without and with geofoam inclusion is shown in Fig. 6. A custom designed and developed strong steel box fabrication was used for small-scale tests on buried pipelines model (with 3 MS steel walls and a glass front, shown in Fig. 6). The fabricated strong box consists of a 10 mm thick steel panel from four sides i.e. bottom, back, and side panels. A 12 mm thick glass panel was placed as the front panel to make easier the two dimensional view of the model. Movements occurring in the fine grained soil mass were captured with the help of a digital camera at a fixed time interval through this transparent front glass panel. The strong box was proof tested for its ability before starting of the tests for different soil backfills and different loading strengths. The loading intensity reaches up to the 50 kN no deformations were observed to occur in the steel panels of the strong box. Measurements were done with the help of image acquisition and analysis and deformations were checked between the box before beginning of the test and during the

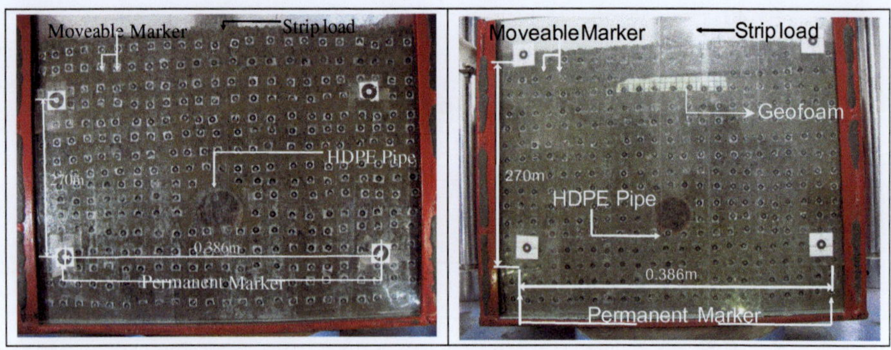

(a) Without geofoam   (b) With geofoam

Fig. 6. Front view of the model test package

tests. The front glass panel was observed to break catastrophically, as the vertical load reaches 50 kN (characteristic value obtained based on 95% successful dummy tests). Inside the strong box numbers of thin polythene sheet strips were placed after application of grease layer. During the test polythene strips were placed such that those moves along with the soil and no boundary friction occur. A fine grained soil was placed at uniform maximum dry density 1.404 gm/cm^3 consistently for the entire test performed. A benchmark was four permanent markers, which were glued to the glass panel so as to measure movements of the moveable markers throughout the advancement of the tests. 'L' shaped plastic markers were in the soil at specific intervals to supervise the movements during the tests. PVC stand arrangement was used to hold Digital camera to make easier the undistorted supervising of the experimental buried pipeline models. EPS8, EPS16 and EPS24 three different geofoam types with varying width were used in the present study, which were placed exactly at depth equal to width of the footing (B_f) for all the model tests. Throughout the progress of the test two sets of lithium battery operated LED lighting panels were used to maintain a constant intensity of illumination.

4.2 Test Procedure

A footing was placed at the center of the test model having 0.203 H wide. Every test of buried pipelines models were tested under a UTM (Universal Testing Machine) at DIEMS Aurangabad with a maximum compressive and tensile capacity of 1000 kN. Load was applied in vertical direction gradually at a constant strain rate of 1.0 N/sec till the maximum settlement of 30 mm reaches or the maximum load of 30 kN reaches (whichever occurs first). Continues sequential images were taken using a Digital camera (Canon make, 9 megapixel, and enhanced shutter speed) with constant time interval equal to two second per image and at the fixed distance from the model such that the picture frame of the camera captures the full view of the model. A connected computer stored the all images throughout the progress of the test and a few meters away from the test setup computer located.

5   Test Program

The details of the model tests performed in the present study shows in Table 2. In the study total 10 model tests were performed with and without geofoam inclusion below strip footing. Model BP11 was tested without any geofoam inclusion and it was treated as the base models for evaluation of the efficiency of the geofoam in reduction of pressure on buried pipelines in cohesive soils.

Table 2. Details of the model tests performed in the present study

Test legend	Geofoam width (mm)	Geofoam density (kg/m^3)
BP11	*N.A	*N.A
BP12	50	8
BP13	100	
BP14	150	
BP15	50	16
BP16	100	
BP17	150	
BP18	50	24
BP19	100	
BP20	150	

*Not applicable as test was performed without geofoam inclusion.

6   Analysis and Interpretation

With the help of open source software package ImageJ, deformations and strains were calculated based on particle movements in sequential images. ImageJ provides analysis and interpretation modules on a set of images. A wide range of measurements are possible on sequential images with software ImageJ. Different macros can be written to formulate a repetitive analysis on the particle movements in the images. With the help of comparative successive analysis in consecutive images, a reference measurement could be made at various points in the images. Using a Template matching, ROI (Region of interest), PIV (Particle image velocimetry) analysis all particle movements could be tracked with incremental images with progress in the test. By using the advanced Template matching and PIV (Particle Image Velocimetry) the displacements occurred in buried pipe was depicted. The deformed profile of buried pipe with surrounding cohesive soil with and without geofoam for strip loading represented in Fig. 7.

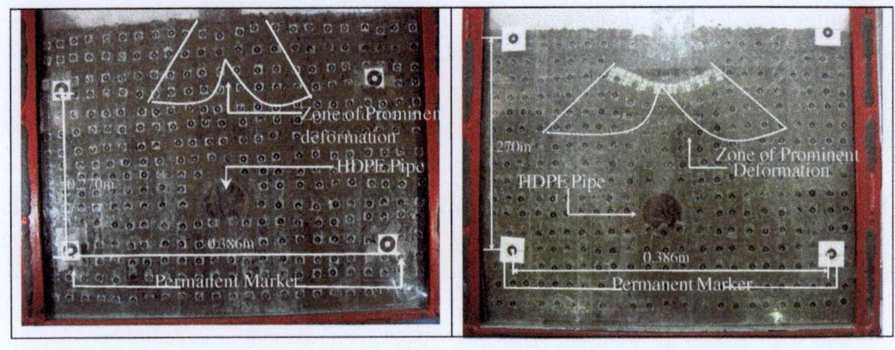

(a) Without geofoam   (b) With geofoam

Fig. 7. Deformed profile of buried pipe and surrounding soil

7   Image Analysis

The open source software ImageJ worked on the images analysis which obtained from the test performed. The advanced template matching plugins and PIV (Particle Image Velocimetry is a velocity-measuring technique. The technique was originally implemented using double-flash photography of a seeded flow. The resulting photographs contain image pairs of each seed particle. For PIV analysis, the photograph is divided into a grid of test patches. The displacement vector of each patch during the interval between the flashes is found by locating the peak of the autocorrelation function of each patch. The peak in the autocorrelation function indicates that the two images of each seeding particle captured during the flashes are overlying each other. The correlation offset is equal to the displacement vector.) analysis are used to measure displacements, which was occurred in, above and around the pipe and geofoam. The deformed view of observational models with and without geofoam is shown in Fig. 7. With the help of image analysis throughout sequence order of images the markers were carried. In this experiment the reference markers were non-displaced and so were used to establish a benchmark for image calibration as well as to measure displacement of the movable plastic markers. When geofoam was included below the shallow, foundation deformations were observed to decrease importantly. When geofoam was introduced in the experiment geometry, the primary deforming zone was found to be concentrating in and around the foundation area. Further, these deformations were observed to be decreasing the seeable heaves at the surface level as well as the deformations in the buried pipe.

The displacement vector diagrams for buried pipeline experimental models without and with geofoam shown in Fig. 8. Comparison is made between two identical models with and without geofoam at a maximum footing settlement of 30 mm. For the improved visualization of the results the vectors in figure are scaled up two times than the original. Punching shear failure could clearly be observed when no geofoam inclusion was made. The fine grained soil deformation carry further to the buried pipe and the zone of plastic equilibrium moves away and forms a heave on both side of the

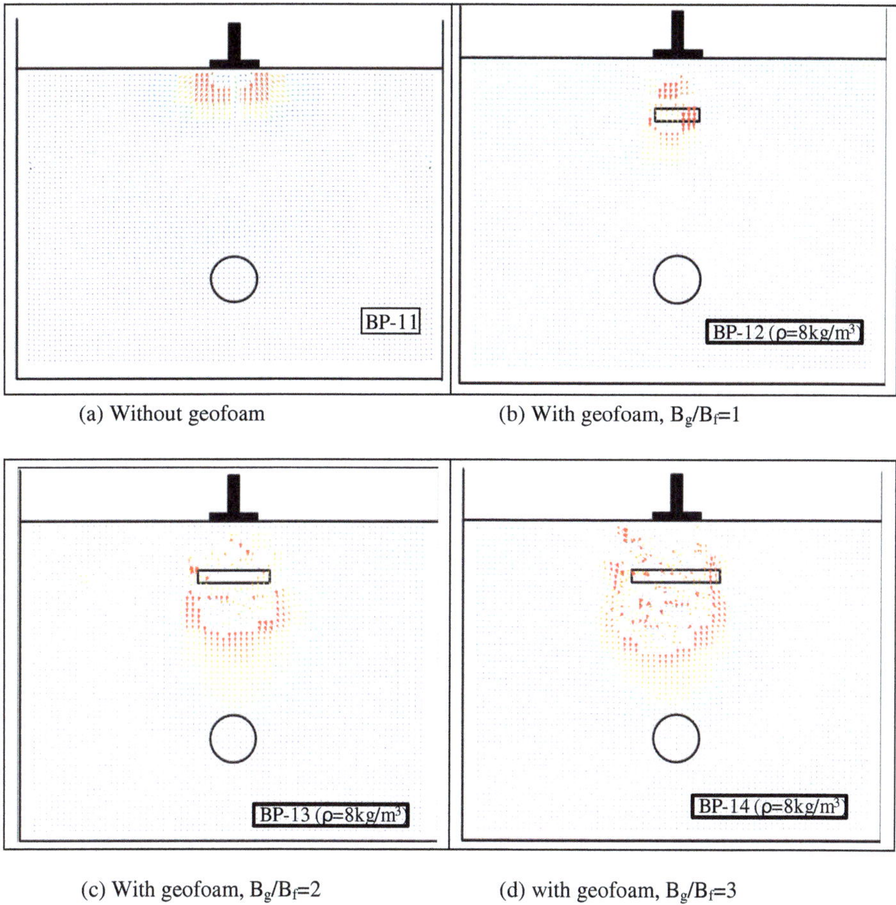

(a) Without geofoam

(b) With geofoam, $B_g/B_f=1$

(c) With geofoam, $B_g/B_f=2$

(d) with geofoam, $B_g/B_f=3$

Fig. 8. Displacement vector for test models with and without geofoam

footing. It observed to decrease with increase in width of the geofoam when geofoam was placed below footing at depth B_f. When the geofoam inclusion provides a compressible bed below footing which compress allowing the load distributed forming an invert arch and load maximum at center and minimum at corner. Due to this the load transfer from axial direction to the outward diagonal directions. Majority of the movements in cohesive soil occurs well above the buried pipe at the same time. This made easier to the shear strength improvement of the cohesive soil and thus transfers reducible loads on the buried pipe.

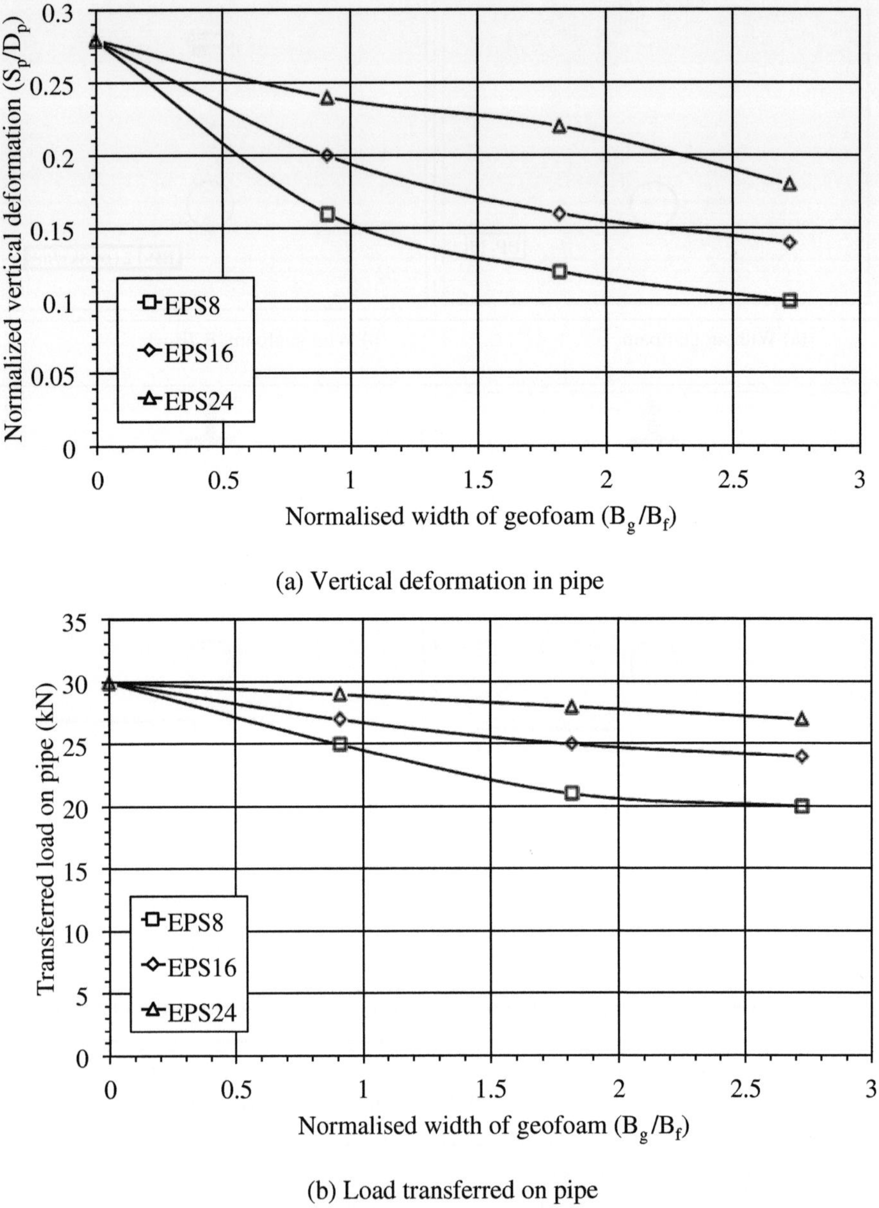

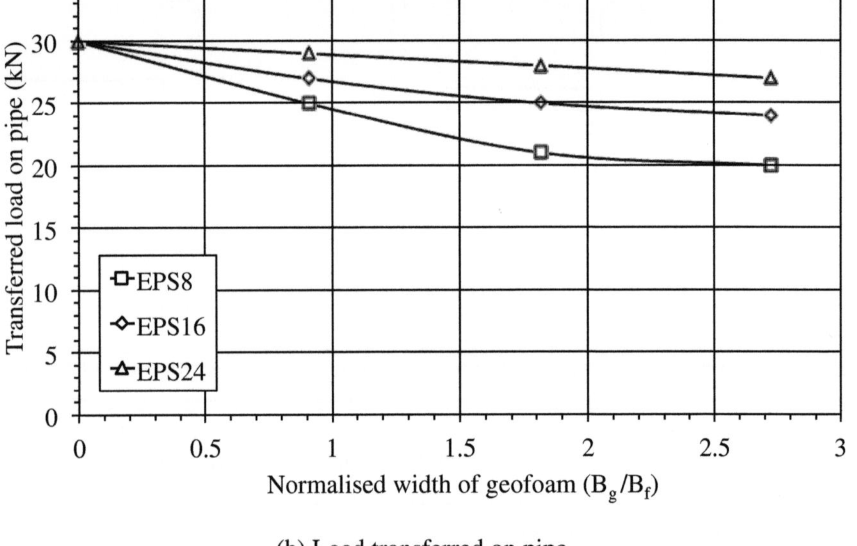

Fig. 9. Influence of geofoam density and thickness on deformation and load transfer

8 Results and Discussion

8.1 *Influence of geofoam width and geofoam density*

Deformations in the vertical direction were calculated from image analysis for all the model tests performed in the present study. It was observed that the deformations in the pipe are inversely proportional to the width of the geofoam and directly proportional to the density of the geofoam. Figure 9(a) shows the variation of vertical deformation occurred in pipe with respect to the normalized width of the geofoam for varying density of the geofoam. A similar representation about the load transferred on the pipe could be made as shown in Fig. 9(b). The load transferred to the pipe was estimated based on the deformations measured through image analysis and the corresponding load from the load deformation diagram of HDPE pipe as shown in Fig. 5. The vertical axis is normalized to the pipe diameter in Fig. 9(a). A maximum reduction in load transferred of up to 32.14% was observed in case of low-density geofoam (EPS8) having maximum width of 150 mm. Table 3 summarizes the results obtained from the series of the model tests performed in this study.

Table 3. Summary of the model tests performed in the present study

Test legend	Geofoam width (mm)	Geofoam density (kg/m^3)	Vertical deformation in pipe, S_p/D_p	Load transferred on pipe (kN)
BP11	*N.A	*N.A	0.024	28
BP12	50	8	0.012	23
BP13	100		0.01	20
BP14	150		0.009	19
BP15	50	16	0.018	26
BP16	100		0.014	23
BP17	150		0.011	21
BP18	50	24	0.022	27
BP19	100		0.018	24
BP20	150		0.014	22

*Not applicable as test was performed without geofoam inclusion.

9 Conclusions

Based on the observations made in the present study, Conclusions made are as below,

1. Geofoam as a compressible inclusion placed below strip footing provides significant reduction in transferred load on buried pipes.
2. As the density of the geofoam inclusion decreases, the load on buried pipes reduces. Which means the load transferred on the buried pipes is inversely proportional to the density of the geofoam.

3. With increase in width of the geofoam load reduction increase. This is mainly due to the load dispersion over wider area in clay above the pipe. The strength of the soil mobilizes with increase in width of the geofoam. A maximum decrease of 32.14% in load on the buried pipe was obtained when a low-density (i.e. 8 kg/m^3) geofoam with maximum width of $3B_f$ was used.
4. So, A maximum reduction in load transfer on buried pipes due to surface loading can be achieved with help of low density geofoam having wider width as a compressible inclusion over the buried pipes.

9.1 *Limitations*

Pertaining to the fact that the small-scale modeling is associated with various limitations following limitations are described in the context of the present study.

1. Small-scale modeling does not induce the identical stress strain conditions as that of field conditions, so the results obtained and presented in this paper should be used only to understand the patterns of load distribution. To interpret the actual analysis and design values a field study or centrifuge model study is recommended.
2. Load transfer mechanism may also be the function of pipe material type and surrounding clay, so a detailed parametric study using this variable is needed to perform to arrive up to suitable implementation of geofoam in the field applications of pressure reduction.
3. The load estimate on the buried pipe was made based on isolate compression test only, however the estimates of the loads must be made based on the confined behavior of the pipe embedded in soil. So this is included in the future scope of the present study.

Acknowledgements. Author would like to thank the reviewers to invest their time and provided valuable suggestions on the improvement of the present manuscript.

References

De Anirban, A.N.M., Zimmie, T.F.: Numerical and physical modeling of geofoam barriers as protection against effects of surface blast on underground tunnels. Geotext. Geomembr. Geotext. Geomembr. **1**, 1–12 (2016)

Bilgin, O., Stewart, H.E.: Studying buried pipeline behavior using physical and numerical modeling. In: GeoCongress 2012: State of the Art and Practice in Geotechnical Engineering, vol. 1, pp. 2128–2137, Oakland, California, United States (2012)

Corey, R., Han, J., Khatri, D.K., Parsons, R.L.: Laboratory study on geosynthetic protection of buried steel reinforced HDPE pipes from static loading. J. Geotech. Geoenvironmental Eng. **1** (1), 1–10 (2014)

Johnson, J., Hutson, A.C., Gibson, R.L., Verreault, L.: Protecting existing PCCP subject to external transient loads. In: Pipelines 2010: Climbing New Peaks to Infrastructure Reliability—Renew, Rehab, and Reinvest, vol. 1, pp. 203–210. ASCE, Keystone (2010)

Lin, T.J., Chou, C.H.: Verification of numerical modeling in buried pipelines under large fault movements by small-scale experiments. In: Fifteenth World Conference on Earthquake Engineering, vol. 1, pp. 1–9, Lisbon, Portugal (2012)

Stephen, S.: Contribution of lateral earth pressure resistance to restrain horizontal thrust in buried pipelines. In: Pipelines-A Sound Conduit for Sharing Solutions, vol. 1, pp. 358–371. ASCE, Carlsbad (2011)

Watkins, R.K.: Pipe and soil mechanics for buried corrugated HDPE pipe. In: Pipelines 2004: Pipeline Engineering and Construction, vol. 1, pp. 1–10. ASCE, San Diego (2004)

Response of Strip Footing Adjacent to Nonyielding Basement Wall

Magdi El-Emam[(✉)] and Majid Touqan

American University of Sharjah, Sharjah, UAE
{melemam, b00035193}@aus.edu

Abstract. A series of 1/3-scale experimental model tests were performed at the American University of Sharjah (AUS) to assess the performance of strip footing constructed adjacent to nonyielding basement wall. Different construction design parameters have been considered including the strip footing width (B), its distance from nonyielding wall back (a), and footing embedment depth below the backfill surface (D_f). The effects of nonyielding wall proximity to the strip footing were investigated. The experimental results show that the lateral deflection of the nonyielding wall is a major factor in dictating the load carrying capacity of the strip footing. The vertically loaded strip footing imposed significant vertical and horizontal forces at the wall top and bottom boundaries. The strip footing load carrying capacity increased as the footing width decreased, the distance from the wall increased and the embedment depth increased. Meanwhile, the vertically loaded strip footing imposed significant vertical and horizontal forces at the wall top and bottom boundaries. The data gathered from this program is used to identify deficiencies in current design methodologies for similar soil structure interaction problems.

1 Introduction

The urgent needs to parking spaces all over the world forced the designers to use multi-levels basement as a viable solution to the parking problems. With the extensive use of multi levels basements the possibility of constructing shallow foundation in the vicinity of old (existing) basement walls increased. The existence of shallow foundations closer to basement walls is expected to impose external lateral pressure in addition to the currently considered lateral earth pressure due to soil. A literature review showed that basement walls are usually designed to resist the soil static lateral earth pressure. This earth pressure is traditionally calculated using the at rest earth pressure theory proposed by Jaky (1944), which proven to be unsatisfactory for heavily compacted soil (Mayne and Kulhawy 1982; Schnaid and Houlsby 1991; Fang et al. 2007; El-Emam 2011).

The additional lateral earth force imposed by shallow foundation constructed in the vicinity of basement walls may cause excessive lateral deflection, damage, or may be failure of the wall. In the meantime, the performance of the shallow foundation may be significantly affected by the performance basement wall (i.e. wall deflection or movement). The design guidelines assume that the footing performance is not affected by the existence of the basement wall, and consider a full bearing capacity for the

© Springer International Publishing AG 2018

M. Bouassida and M.A. Meguid (eds.), *Ground Improvement and Earth Structures*,
Sustainable Civil Infrastructures, DOI 10.1007/978-3-319-63889-8_3

footing. In the meantime, for the basement wall design the method only considers the lateral earth force resulted from the strip footing. The lateral earth pressure imposed on vertical basement wall by strip surcharge load is usually calculated using the modified Boussinesq's Equation (Jarquio 1981; Misra 1981; Barnes 2010) is based on theory of elasticity. Assuming linearly elastic behavior of this problem is expected to lead to neglecting the elastic nonlinearity and the plastic responses of the footing and the deformed wall. Plastic responses of both soil and basement wall are expected when the loading intensity on the footing is large and/or the wall rigidity is small. Assuming elastic behavior of soil might be resulted in underestimation of lateral earth forces calculated using Boussinesq's Equation, and in turn lead to unsafe design of the wall. In addition, this method neglects effect of soil strength on lateral-earth pressure against a wall, which is proven to be a major factor, Misra 1981 and Motta 1994. Therefore, a comprehensive analysis and design method that consider the movement and rigidity of the wall, and elastic and non-elastic responses of wall-soil-footing system is needed. The current research is a major step towards the development of this method of analysis.

Previous research work has indicated similar needs for the comprehensive analysis and design theory. Rehnman and Broms (1972) conducted a series of full scale model tests of reinforced concrete retaining walls to investigate the compaction induced stresses. In the experimental models two types of backfill were used; gravelly sand and silty fine sand. The loading of the experimental models simulated the application of wheel loads as surcharge of a pair of wheels with 2 m apart, 7.5-ton loads and 1 m away from the wall. Results showed that the measured peak pressure increased during loading, and the residual pressure increased after removal of point surcharge load. Dave and Dasaka (2012) examined both magnitude and distribution of earth pressure with reference to rigid cantilever retaining wall movements. They concluded that the earth pressure due to surcharge loading was greater near the top of the wall and decreased nonlinearly down the wall. Finally, Dave and Dasaka (2012) concluded that Jaky's equation underestimates the earth pressure up to the mid height of wall, and overestimates it in the remaining section of the wall with depth. Results showed that as the surcharge-wall distance increases, the earth pressure imposed on the wall reduces. Fang et al. (2007) conducted experimental models where a strip footing was placed at distances away from the wall a = 0.15 m, 0.20 m, 0.40 m. It was found out that the experimental lateral soil force induced by surcharge loading was apparently smaller than the theoretical predictions. In addition, they reported that Jaky's solution underestimated the lateral earth pressure in the compaction influenced zones near the top of the wall that have experienced high pressures due to compaction.

The main objective of the current research is to investigate the behavior of strip footing constructed adjacent to nonyielding basement wall. In this paper, results of three reduced-scale wall-strip footing model testes are reported. The emphasis of the paper is on the quantitative effect of the wall on footing load carrying capacity and settlement and the effect of the footing on the wall responses. The strip footing distance from the wall is the major parameter that is represented in this paper.

2 Laboratory Testing Program

2.1 Testing Facility and Materials

A total of fifteen 1.15 m-high model nonyielding walls-strip footing system constructed at 1/3 scale comprise the experimental portion of the current research program. Results of three model walls from this series of tests are presented that demonstrate the influence of the strip footing width (B) on the response of nonyielding wall-footing to applied vertical stresses. Figure 1 shows a typical nonyielding wall-strip footing model test. The wall thickness was 7.5 cm, the footing width was selected to be B = 15 cm and 25 cm, and the distance between the footing and the wall was selected to be a = 3B. The model walls were designed in accordance with similitude rules proposed by Iai (1989), to ensure that the geometry of the walls, soil properties and strip footing parameters were typical for a prototype (i.e. field scale). The similitude laws proposed by Iai (1989) using the geometrical scale factor $\lambda = 3$ in this investigation, where $\lambda = H_p/H_m$, and H_p and H_m are the heights of both prototype and model, respectively. More details about Iai's similitude could be found at El-Emam and Bathurst (2004, 2005, 2007) who successfully adopted the same scale factors in producing 1-g shaking table models for reinforced soil retaining wall.

Fig. 1. 1/3 Nonyielding wall-strip footing model under construction with full height panel facing.

The backfill material used in the model tests was acquired from Ras Al-Khaimah Quarry, UAE. According to the Unified Soil Classification System (USCS), the sand can be classified as well graded sand (SW), with angular to sub-angular particles. This sand was selected because it is locally available and used with many of local construction projects. The sand had a maximum dry unit weight of 17.9 kN/m^3 at Optimum Moisture Content of 5.5%. Direct shear tests on the sand prepared to the same unit weight gave a peak direct shear friction angle $\phi = 43°$. Figure 2 indicates the grain size distribution for RAK sand used in this study.

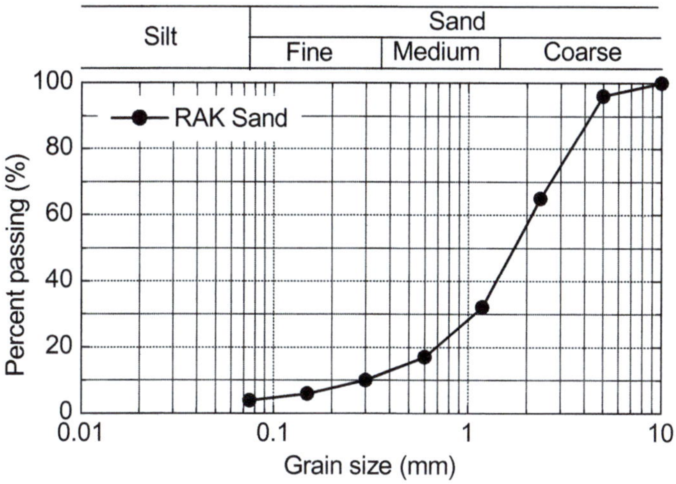

Fig. 2. Grain size distribution curve for Ras Al-Khaimah sand.

The wall facing was constructed from 23 hollow steel box-sections bolted together with 3–20 mm diameter rods to form a 1.15-m high rigid panel. The strip footing was made of rigid wood plate of 136 cm length (representing the width of the steel box) and different width B = 15, 20, and 25 cm. The wooden plate is reinforced with two-10 cm by 10 cm cross section steel beams welded together. A sand layer was glued to the bottom face of the wooden plate (i.e. the face in contact with the soil) in an effort to reduce slippage between the footing and the sand backfill (see Fig. 1).

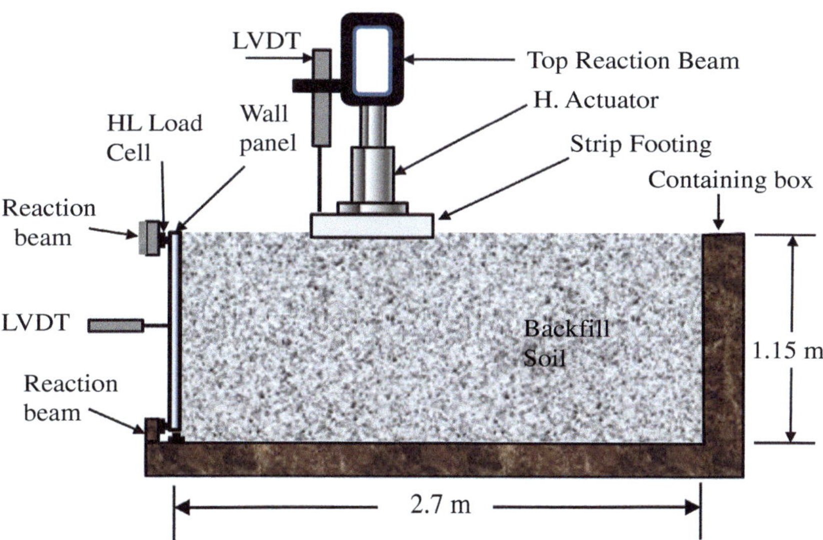

Fig. 3. Schematic for nonyielding wall-strip footing model with different types of instrumentations.

2.2   Instrumentations and Applied Vertical Stresses

Up to 16 pieces of instrumentation were deployed in each model wall-footing system. The wall horizontal deflection was measured using a Linear Voltage Displacement Transducer (LVDT) that was mounted against the mid height of the wall.

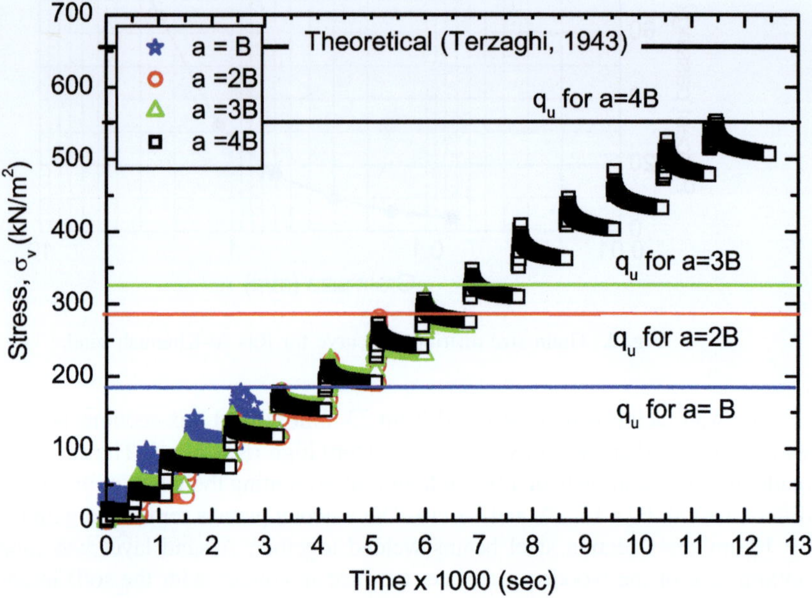

a)   Footing vertical stress

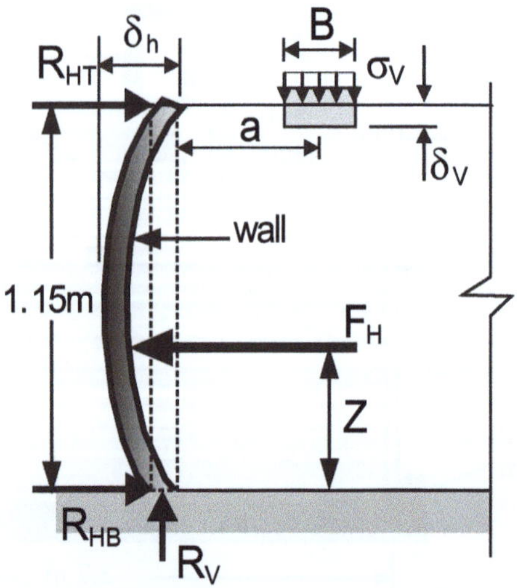

b)   Schematic of the wall-footing responses

Fig. 4. Footing vertical stress increments versus time for three model tests, and schematics of the responses measured during the tests.

Three vibrating wire transducers pressure cells were used to measure the lateral soil pressure at the back of the wall during loading stages. Two vertical and two horizontal load cells were installed at the base of the wall panel to measure the forces transmitted to the footing (wall toe) both vertical and horizontal, respectively. In addition two-horizontal load cells were installed at the top of the wall panel to measure the horizontal force transmitted to the top. Figure 3 shows a schematic for strip footing-nonyielding wall with different types of instrumentations.

The strip footing was instrumented with two load cells, inserted between the footing and the hydraulic actuator, to directly measure the vertical load applied at the footing. In addition, three LVDTs were attached to a rigid wooden beam above the strip footing to measure the vertical settlements at three different locations of the strip footing (Center, left side and right side). Finally, InstruNet World Software was utilized to collect data from all sensors using the data acquisition and a fast PC.

The vertical stress was applied on the strip footing as stepped-stress increments, as shown in Fig. 4a. Each stress increment ($\Delta\sigma_v$) is held for 15 min before applying the next stress increment (Jalla 1999) (ASTM 1994). The vertical stress increased until the model failed due to large footing vertical settlement (i.e. $\delta_v = 0.1B$) or large wall horizontal deflection or both. Loading sequences and procedures are applied to follow the Plate Load Test (PLT) standards (ASTM 1994). In some cases tests were ended when the ratio of load increment to settlement increment reached minimum. (ASTM 1994) In other cases of excessive horizontal wall deflection the tests were stopped to maintain the integrity of the nonyielding wall panel.

3   Results and Discussion

3.1   Strip Footing Vertical Stress and Settlement

3.1.1   Footing Vertical Stress

In this paper, time histories of the measured responses for three model tests are presented as typical results of the measured responses of all 15 model tests. The three models were constructed with footing width 25 cm. The strip footing is constructed at distances from the wall a = 1B, 2B, 3B and 4B. To have a uniform base of comparison, the applied vertical stresses for each footing versus time increment is shown in Fig. 4a. It is clear that there is perfect agreement between the three stress histories after the first 3 increments. At the first 3 stress increments, a slight deviation in the applied stresses could be noticed. Figure 2a indicated that, stress applied on footing of model with footing distance a = B was slightly larger compared to stress applied for the models with footing constructed at other distances (a). This stress difference might be taking into consideration in the following discussion, as the comparisons will be in the time domain. Whenever the comparison is in stress domain, the effect of vertical stress difference will be eliminated. It should be noted that the wall with a footing distance a = 1B reported to be reach failure at the fourth vertical stress increment.

3.1.2   Footing Vertical Settlement

Footing vertical settlement increment for the three model tests versus time are shown in Fig. 5. It is clear that the footing vertical settlement increased as the applied vertical stress increment increased for all model tests. In addition, for the same vertical stress increment, the model with smaller footing distance (a) showed larger vertical settlement except for mode wall with footing distance a = 1B. This model (i.e. with a = 1B) showed less vertical footing settlement compared to the other three models in the first three time intervals (up to t = 2500 s), despite its larger stress increments at these time intervals. This behavior is attributed to the nature of footing soil interactions during loading process, shown in Fig. 6. As the strip footing constructed adjacent to the nonyielding wall (a = B), the soil carrying the footing acts like a column, Fig. 6a. Therefore, in this case the engaged soil properties are the confined compressive strength rather than shear strength. As the strip footing constructed further from the nonyielding wall (i.e. a > B), a failure wedge shown in Fig. 6b generated, and the soil behaves in shear strength. For these reasons the footing with a = b showed less settlement at the first 3 stress increments. Once the nonyielding wall started to deflect laterally, the soil column in Fig. 6a fails in both strength loss and buckling action. This caused the footing to suffer dramatic increase in the vertical settlement, as shown in Fig. 5a for a = B. For other footings constructed at distances a > B, the soil wedge engaged in carrying the footing increased with the footing stress, and therefore, resistance to failure increased. This prevents footings from suffering large vertical settlement. Actual settlement of a typical footing at the end of loading stages is shown in the picture of Fig. 5b.

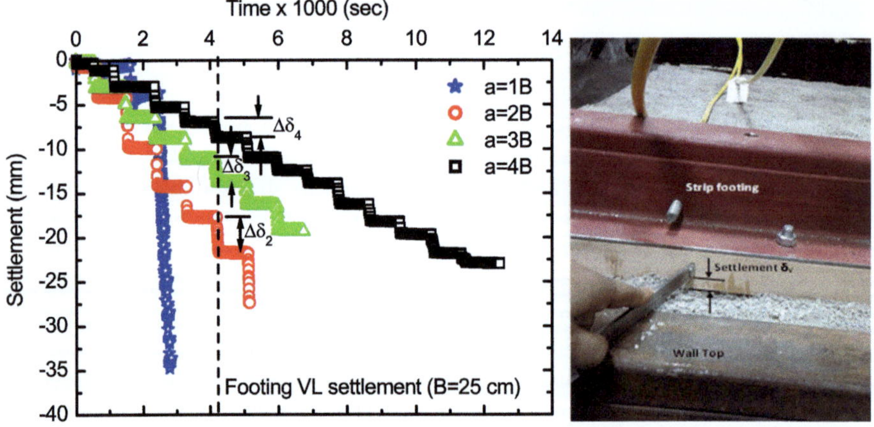

Fig. 5. Footing vertical settlement increments versus time for the three model tests.

Another important behavior is the vertical settlement increment ($\Delta\delta$) associated with same stress increment at the footing constructed at different distances from the wall (a). Figure 5a indicated that the incremental vertical settlement at the same stress increment decreased as the footing distance from the wall (a) increased. A close look at

Fig. 5a shows that the vertical strain increments $\Delta\delta_2 = 3.9$ mm, $\Delta\delta_3 = 2.5$ mm and $\Delta\delta_4 = 1.8$ mm, for footing constructed at distances a = 2B, 3B and 4B respectively. This is occurred at elapsed time t = 4200 s and at vertical stress increment $\Delta\sigma_v = 66$ kN/m^2.

3.2   Lateral Deflection of Nonyielding Wall

Figure 7 shows the lateral deflection (δ_h) measured at the med-height of the nonyielding wall versus time for model with different footing distance (a). For all models the wall lateral deflection (δ_h) increased as the vertical stress over strip footing increased. This is expected and attributed to the larger horizontal forces developed at the back of the wall supporting the strip footing with smaller distance (a) (see Fig. 9). In addition, the strip footing with smallest distance to the wall (i.e. a = B) caused extremely larger lateral deflection on the wall, despite the near equal vertical stress increments, compared to other models. This is explained in term of buckling failure that occurred for the soil column under the footing (Fig. 6a) due to a slight lateral deflection of the wall adjacent to the soil column. In this case (a B), the closeness of the footing to the wall does not give a chance to the soil shear wedge to formulate. Finally, the application of the fourth stress increment on the strip footing caused a lateral deflection increments, $\Delta\delta_h = 24.3$ mm, 2.98 mm, 0.98 mm, and 0.73 mm for models with footings constructed at distances a = B, 2B, 3B and 4B respectively. This lateral deflection reflected largely on the bearing capacity of strip footing as shown in Fig. 4a.

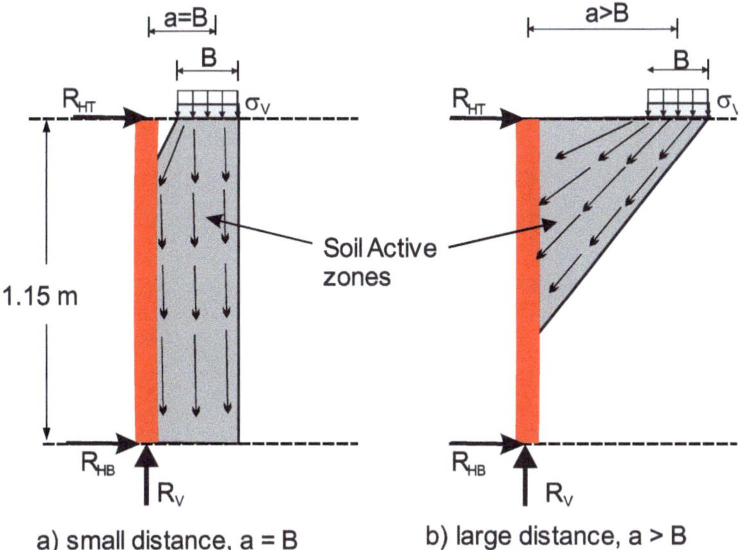

Fig. 6. Footing-soil interaction for footings constructed at small and large distances (a).

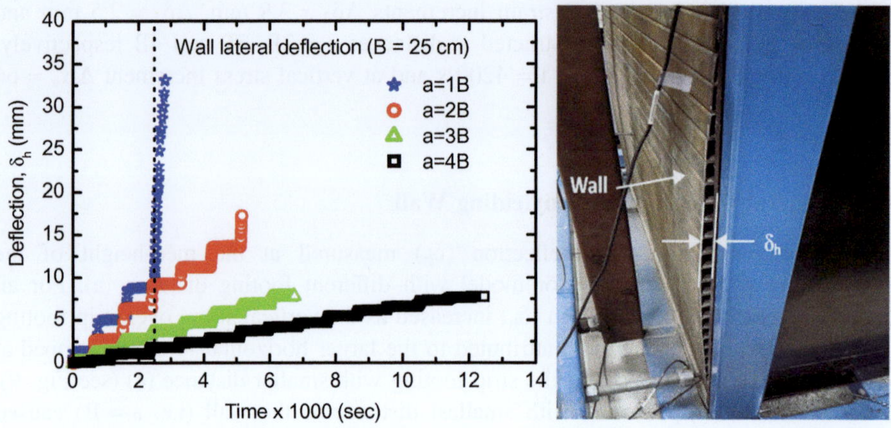

Fig. 7. Lateral deflection of the wall (δ_h) versus time for models with footings constructed at different distances from the wall (a).

3.3   Force Response of Nonyielding Wall

3.3.1   Wall Footing Vertical Forces

Vertical forces (R_V) acting at the bottom (foundation) of the nonyielding wall are shown in Fig. 8 for model walls with different footing distances (a). This load is developed due to the application of the vertical stress at the top of the strip footing, and

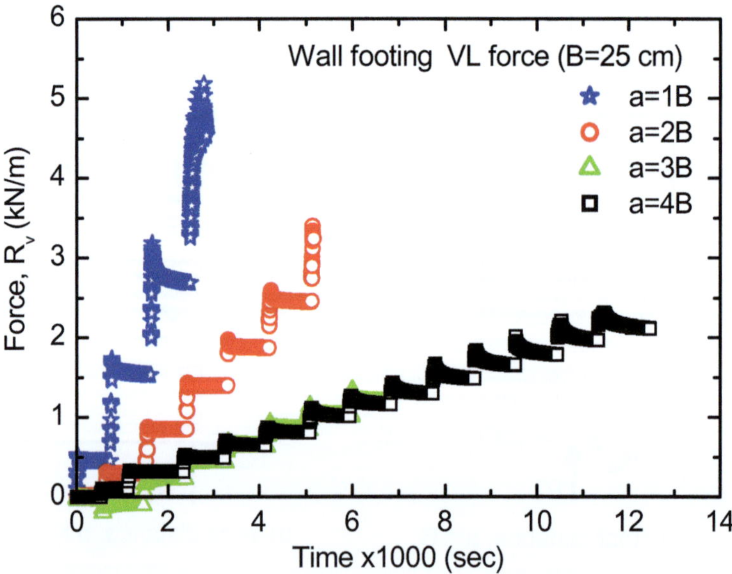

Fig. 8. Vertical earth forces at the back of the wall due to strip footing vertical stresses at different distances from the wall (a).

not including the self-weight of the wall (i.e. W_w = 1.3 kN/m). At the same vertical stress, the footing with smaller distance to the wall (i.e. a = B) imposed larger vertical force at the wall footing. As the distance a increased, the vertical force imposed by the footing on the wall decreased significantly. This could be attributed to the volume of soil affected by the strip footing (i.e. the size of developed soil wedge). It is well understood that the volume of soil affected by the footing (soil wedge in Fig. 6b) increased with the footing distance from the wall (a). As a result the surface which resists the soil failure increased, leading to a decrease in the vertical force R_V. It should be noted that the force R_V needs to be considered in the design of the wall foundation, otherwise the design will be unsafe.

3.3.2   Wall Horizontal Forces

Horizontal forces (F_H) developed at the back of the nonyielding wall are shown in Fig. 9 for models with strip footings constructed at different distances (a). These forces are developed due to the application of vertical stresses at the top of the strip footings, and in addition to the at rest lateral earth pressure resulted from the soil itself. This force was measured using load cells attached to the top and bottom boundaries of the wall (Fig. 3). It is very obvious that the strip footing with larger distance from the wall (a) imposed less horizontal force (F_H) at the back of the wall. This is due to the larger resisting soil wedges associated with larger footing distances. As the soil wedge is getting larger, it resists the footing stress more, and reduces the lateral force transferred to the wall. This reduces the lateral pressure at the back of the wall, and therefore

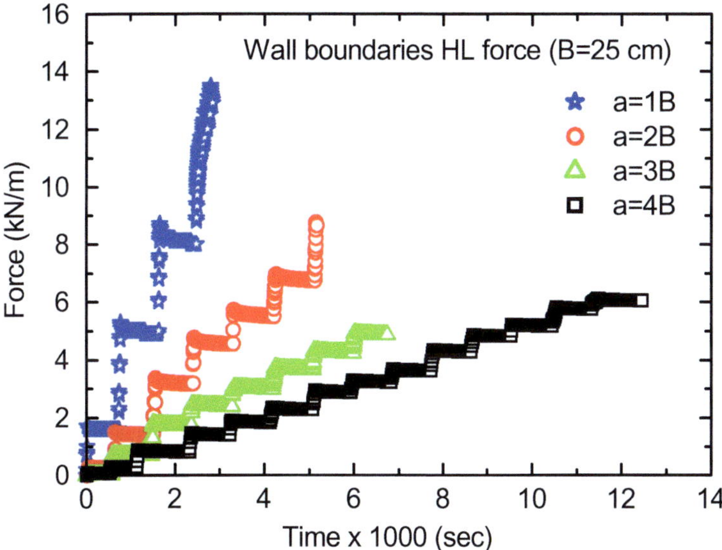

Fig. 9. Lateral earth forces at the back of the wall due to strip footing vertical stresses at different distances from the wall (a).

reduces the horizontal forces F_H. The incremental lateral force (ΔF_H) at the same stress increment is also decreased largely as the footing distance from the wall (a) increased. This forces should be considered in designing the footing of the wall and the concrete slab of the first basement.

4 Conclusions

A large number of wall-foundation systems were experimentally analyzed in this study. The results were presented in the form of foundation load settlement curve for different footing width. Also, the wall lateral forces and deflection for different models were presented and discussed. Some important conclusions and implications to current design practice are summarized below.

- Footing vertical settlement increased as the applied vertical stress increment increased for all model tests. For the same vertical stress increment, the strip footing vertical settlement increases as the footing distance to the wall (a) decreases.
- Strip footing ultimate bearing capacity increases as the footing distance to the wall (a) increases. For footing with distances a < 3B from the wall, the ultimate bearing capacity is expected to be less than Terzaghi ultimate bearing capacity.
- Footing with smaller distance to the wall (a) imposed larger vertical force at the wall footing (R_V). Neglecting this additional vertical force in the current code of practice might lead to unsafe design output.
- Strip footing with smaller distance to the wall (a) imposed larger horizontal force (F_H) at the back of the wall, despite the near equal vertical stress (σ_V) applied at the both strip footings. This force should be practically considered in designing the wall.
- The lateral deflection of nonyielding wall increased as the vertical stress over the strip footing (σ_v) increased, and as the strep footing distance to the wall (a) decreased.

Acknowledgments. The first author is grateful for funding provided by the office of research and graduate studies (ORGS) at the American University of Sharjah (AUS), UAE (FRG14-2-19).

References

ASTM: Standard Test Method for Bearing Capacity of Soil for Static Load and Spread Footings. ASTM, West Conshohocken (1994)

Barnes, G.: Soil Mechanics, Principles and Practice, 3rd edn. Palgrave Macmillan Publisher, Basingstoke (2010)

Dave, T., Dasaka, S.: Transition of earth pressure on rigid retaining walls subjected to surcharge loading. Int. J. Geotech. Eng. **2012**(6), 427–435 (2012)

El-Emam, M.M., Bathurst, R.J.: Influence of reinforcement parameters on the seismic response of reduced-scale reinforced soil retaining walls. Geotext. Geomembr. **25**(1), 33–49 (2007)

El-Emam, M.M., Bathurst, R.J.: Experimental design, instrumentation and interpretation of reinforced soil wall response using a shaking table. Int. J. Phys. Model. Geotech. **4**(4), 13–32 (2004)

El-Emam, M.M., Bathurst, R.J.: Facing contribution to seismic response of reduced-scale reinforced soil walls. Geosynth. Int. **12**(5), 215–238 (2005)

El-Emam, M.: Experimental and numerical study of at-rest lateral earth pressure of over-consolidated sand. Adv. Civ. Eng. **2011**, 12 p (2011)

Fang, Y., et al.: Earth pressure on an unyielding wall due to a strip surcharge. In: International Offshore and Polar Engineering Conference, pp. 1233–1238 (2007). Lisbon

Iai, S.: Similitude for shaking table tests on soil-structure-fluid model in 1g gravitational field. Soils Found. **29**(1), 105–118 (1989)

Jaky, J.: The coefficient of earth pressure at-rest. J. Soc. Hungarian Arch. Eng. **78**, 355–358 (1944). (Hungarian)

Jalla, R.: Basement wall design: geotechnical aspects. J. Arch. Eng. **5**, 89–94 (1999)

Jarquio, R.: Total lateral surcharge pressure due to strip load. Geotech. Eng. Div. ASCE **107**(10), 1424–1428 (1981)

Mayne, P., Kulhawy, F.: Ko-OCR relationships in soil. J. Geotech. Eng. Division ASCE **108** (GT6), 851–872 (1982)

Misra, B.: Lateral pressures on retaining walls due to loads of surface of granular backfill. Soils Found. **20**(2), 31–44 (1981)

Motta, E.: Sulla valutazione della spinta attiva in terrapieni di altezza finite. Rivista Italiana di Geotecnicano. **XXVII**(3), 235–245 (1994). (in Italian)

Rehnman, S., Broms, B.: Lateral pressures on basement wall. Results from full-scale tests. In: 5th European Conference on Soil Mechanics and Foundation Engineering, pp. 189–197 (1972)

Schnaid, F., Houlsby, G.T.: Measurement of the properties of sand by the cohesion pressuremeter test. University of Oxford, Oxford, Soil mechanics Report 113/91 (1991)

3D Modeling of EPS Geofoam Buffers Behind Diaphragm Walls

Salem A. Azzam[✉], Beshoy M. Shokry, and Sherif S. AbdelSalam

Civil Engineering Department, Faculty of Engineering,
The British University in Egypt, Cairo-Suez Road, Al-Sherouk 11837, Egypt
{salem.azzam, beshoy115377,
sherif.abdelsalam}@bue.edu.eg

Abstract. In an attempt to reduce lateral earth pressures acting on diaphragm walls, decrease the dependency on anchors, and optimize the wall structural design, expanded polystyrene geofoam (EPS) was introduced as a compressible buffer between the wall and the retained soil. Based on verified outcomes from the literature, EPS buffers is an effective solution that can significantly reduce the static lateral earth pressure acting on flexible walls. In this paper, a 3D numerical model was developed for a small-size diaphragm wall with EPS buffer using the finite element (FE) program PLAXIS 3D. The constitutive properties utilized in the model were measured as part of the material characterization phase of this research project, and the model was intended to capture the short-term behavior of the retained soil using EPS buffers with various thicknesses. To verify the FE results, a physical instrumented prototype was assembled to mimic the modeled diaphragm wall with EPS. The comparison showed a decent agreement between the FE results and the prototype measurements. From the main outcomes, lateral pressure on diaphragm walls was significantly reduced by around 37% using a relatively thin EPS buffer.

Keywords: Geotechnical · Geosynthetics · Geofoam · Diaphragm walls · EPS buffer · Lateral pressure · Modeling

1 EPS Buffers for Retaining Walls

Diaphragm walls are an expensive type of retaining systems compare to cantilever and gravity walls. Many large infrastructure projects require diaphragm walls, as lateral pressure represents the main acting loads that lead sometimes to wall oversized dimensions and may require single or multi-anchors, which makes this type even more expensive. In diaphragm walls, the geotechnical stability is achieved by providing a sufficient wall embedded depth that prevents rotation, while the overall stability must be achieved by conducting a slop stability check. The structural stability of these walls require sufficient resistance against straining actions induced due to the lateral earth pressure, which mainly comes from the soil behind the diaphragm wall.

Expanded polystyrene (EPS) geofoam is a lightweight material that can be used instead of soil for backfill behind retaining walls to reduce the lateral pressures applied on the walls, as the relation between lateral pressure and backfill weight is proportional.

© Springer International Publishing AG 2018
M. Bouassida and M.A. Meguid (eds.), *Ground Improvement and Earth Structures*,
Sustainable Civil Infrastructures, DOI 10.1007/978-3-319-63889-8_4

Therefore, soil backfill behind cantilever walls can be replaced by a material such as foam to achieve static stability (Karpurapu and Bathurst 1992; Horvath 1994; Zarnani and Bathurst 2008; Lutenegger and Ciufetti 2009; Athanasopoulos et al. 2012; Ertugrul and Trandafir 2013; AbdelSalam and Azzam 2016). For flexible walls, limited research was conducted indicating that the earth pressure can also be reduced. Recently, AbdelSalam et al. (2016) have modeled flexible cantilever walls using the three-dimension (3D) finite element (FE) software package PLAXIS 3D Anniversary Edition (Brinkgreve et al. 2015) to identify expected amount and possible reasons of reduction in lateral pressure. It was noticed that an arch shape was evident in the flow of the lateral stresses after using EPS buffer, which is due the relative large compression that occurred within the bottom segment of the EPS, where part of the lateral stresses was carried via friction by neighboring soils.

Diaphragm walls can be considered as flexible embedded wall, hence a reduction in the lateral earth pressure can be achieved behind diaphragms as well, but if an EPS sheet is placed as buffer between the wall and soil behind it. However, most of the current research focused on modeling EPS behind rigid walls, some research focused on flexible walls, and the rest focused on seismic behavior of EPS such as researches done by Athanasopoulos et al. (2012), Padade and Mandal (2014), AbdelSalam and Azzam (2016), and others. There is almost no research that contain information from verified laboratory and/or numerical models to solve diaphragm walls with EPS buffer, and this is because the method of placing EPS between wall and soil is still unknown.

In this paper, the EPS Geofoam was proposed as a thin buffer between the diaphragm wall and the soil behind it, in an attempt to understand the change in the lateral pressure acting on the wall. A method of statement for placing the EPS buffer was proposed. Local EPS properties were summarized based on AbdelSalam et al. (2015) who conducted a series of laboratory tests for EPS characterization. Then, a diaphragm wall was modeled using PLAXIS 3D to identify the expected amount of reduction in lateral pressure behind such flexible wall type, and using various thicknesses of EPS buffers. Finally, the FE results were verified against measurements from a physical prototype constructed as part of this study. The main outcomes included design charts that show the expected overall reduction in the lateral pressure behind diaphragm walls with respect to the EPS thickness.

2   EPS Mechanical Properties

AbdelSalam et al. (2015) have conducted a series of tests to determine the local EPS material properties. The laboratory testing program consisted of unconfined compression (UC) tests to determine stiffness and modulus of elasticity (Young's modulus) for EPS by using triaxial machine. The direct shear test (DST) was used to determine the internal shear strength (internal friction angle and cohesion) between EPS Geofoam beads, and modified direct shear test (mDST) to determine the external shear strength (external friction angle) between EPS blocks and other surface. More than 20 modified DST between EPS and other materials (complete testing procedures are found in AbdelSalam et al. 2015) were conducted to characterize the material and interface properties of local EPS and to serve as basis towards the calibration of the hardening soil constitutive model. Also the reduction factor (Rinter) values for various interface surfaces between EPS and other materials with respect to contact pressure were provided by AbdelSalam and Azzam (2016), who indicated that the relation between Rinter and the contact pressure is non-linear and tends to be constant at high stress levels.

3   Numerical Model

A one-meter embedded wall was simulated in 3D using PLAXIS 3D as shown in Fig. 1a. The standard soil behind the wall with a density around 20 kN/m^3 and the internal friction angle 35° was replaced with the EPS Geofoam with a density of around 20 kg/m^3, using various EPS thicknesses (t) to wall height (h) ration starting from t/h = 0.025, 0.05, 0.09, and 0.13 (where t is the EPS thickness, and h is the wall height). This large range of t/h was used in order to determine the percentage of the reduction in lateral earth pressure acting on the embedded wall with respect to EPS thickness. Figure 1a also shows the deformed mesh after running the model for EPS with ratio t/h = 0.05, and results were extracted for the phase where the excavation depth was 45 cm from the wall top. Figure 1b represents displacement in x-direction (normal to wall plan) with maximum value of 0.15 mm. Figure 1c shows displacement in z-direction (parallel to wall plan) with maximum value of 0.85 mm.

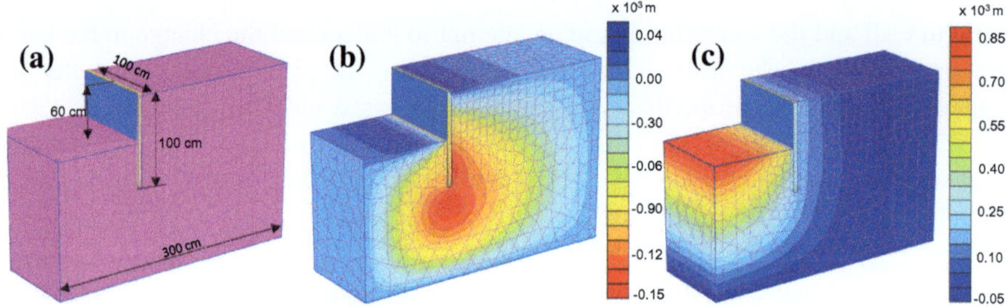

Fig. 1.   Wall with EPS: (a) deformed mesh; (b) disp. in x-dir.; and (c) disp. in z-dir.

The constitutive models that were used to model EPS, concrete, and sand were the hardening soil model, linear elastic, and Mohr-Coulomb, respectively. Properties for the EPS were assumed after Azzam and AbdelSalam (2015). The properties and the constitutive models used for all materials in the FE analysis are included in Table 1.

Figure 2a shows the lateral earth pressure (in x-direction) where the range of the lateral pressure was 6.7 kPa at the toe of the embedded wall without EPS buffer, and was reduced to 5.44 kPa by using EPS buffer with t/h = 0.05. Therefore, the reduction was around 19% by using t/h = 0.05. The FE model with t/h = 0.13 showed more reduction in the lateral stress to reach only 3.5 kPa. Figures 2b and c show the pressure on the wall with t/h = 0.05 in y- and z-directions, respectively.

According to AbdelSalam et al. (2016), a main reason for the reduction of the lateral pressure acting on the diaphragm wall with EPS buffer could be due to the

Table 1. Material properties used in FE model

Material model		EPS hardening	Concrete linear elastic	Sand Mohr-Coulomb
Unite weight	γ (kN/m^3)	0.2	22	18.5
Cohesion	C (kPa)	12	N/A	1
Friction angle	φ (°)	33	N/A	30
Initial stiffness	E_{ref} (kPa)	3004.438	30×10^6	1.3×10^4
Interface factor	R_{inter}	0.7	Rigid	0.67

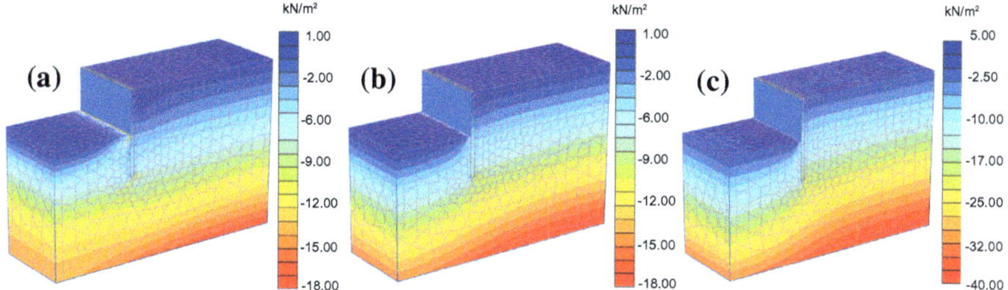

Fig. 2. Stress on wall with t/h = 0.05: (a) x-dir.; (b) y-dir.; and (c) z-dir.

relative large compression that occurred within the bottom segment of the EPS, which caused larger movement within the bottom portion of the soil backfill, leading to the development of shear stresses on the sides of the moved soil portion, and hence part of the lateral stresses was carried by friction with the neighboring soils. Accordingly, deformed mesh, lateral deformation, and contact pressures acting on the +ve interface plate were presented in Fig. 3. The figure shows agreement with the literature, as the maximum deformation and stresses in the case of t/h = 0.05 was still in the bottom third of the wall, and hence reduction in the lateral pressure occurs.

Figure 4a illuminates the effect of changing the EPS thickness on reduction of the lateral earth pressure along the wall depth. From the figure, it was observed that the lateral earth pressure increases nonlinear with wall depth which is expected. The lateral pressure reached a value of 0.33 kPa, while it decreased using EPS with t/h = 0.025, 0.05, 0.09, and 0.13 to reach around 5.70, 5.44, 4.10, and 3.50, respectively. In Fig. 4b,

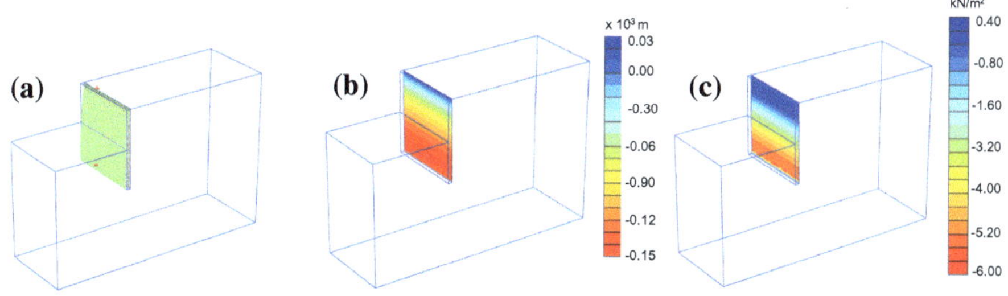

Fig. 3. Wall +vet interface: (a) deformed mesh; (b) lateral deformation; (c) lateral stress

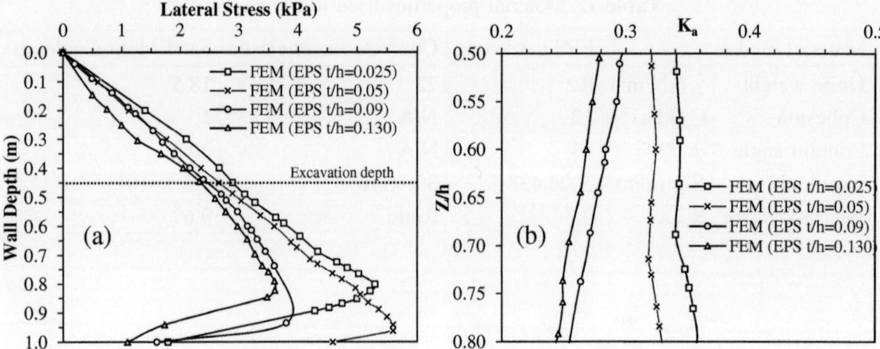

Fig. 4. (a) Reduction in stress versus wall depth; and (b) normalised stress versus depth

the lateral pressures (σ_x) was normalized with respect to the vertical overburden stress (σ_v), which is corresponding to K_a. The elevation (Z) is represented on the vertical axis of the figure. It was measured from the wall top and positive downwards, and was normalized against wall height (h), whereas horizontal axis represents K_a. As can be seen from the figure, the average value of K_a after using the smallest EPS buffer of t/h = 0.025 was around 0.34, whereas this value was reduced to 0.32, 0.28, and 0.26 when using EPS buffers with t/h = 0.05, 0.09, and 0.13, respectively.

4   Verification Using a Physical Prototype

4.1   Prototype Description

A physical prototype was assembled in the laboratory to verify the results of the FE model, with length equal to 140 cm, width 60 cm, and height 100 cm. The side walls at the boundaries of the prototype were made of thick plywood, braced along the perimeter using horizontal struts, and at the corners using four vertical struts. The base of the prototype was directly placed on a concrete slab-on-grade. Figure 5a shows a photo of the prototype wall with EPS buffer fixed along its length, whereas this figure shows the buffer with thickness equal to 25 mm (i.e., t/h = 0.025). The flexible diaphragm wall of the prototype was made of plywood with thickness equal to 12 mm without bracing, so that the stiffness of the wall is small to fully mobilize the active earth pressure. The wall height of the prototype was 100 cm, totally fixed from the bottom with the base, and from the sides with a height of 55 cm from the bottom. These boundary conditions were made based on manual calculations assuming that the wall embedded in the soil and has a free length of 45 cm. More information about the prototype can be found in AbdelSalam et al. (2016).

As shown in Fig. 5b, the soil backfill was initially placed and compacted on both sides of the wall in 5 equal layers of 20 cm, then removed in 10 equal layers (5 cm each) from the passive side. The soil backfill was course sand with same properties summarized in Table 1. Figure 5c shows the passive side after excavation to a maximum depth of 45 cm from the wall top, and also shows the in-house data acquisition

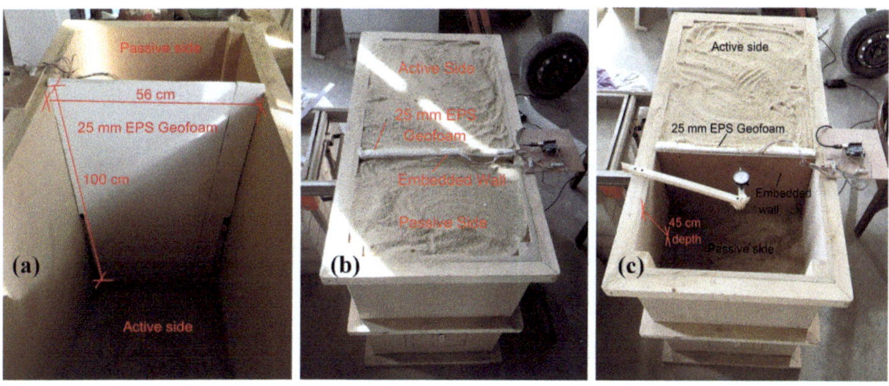

Fig. 5. Prototype (a) wall with EPS buffer; (b) initial stage; and (c) final excavation

board that was connected with four pressure sensors installed along the wall length from the active side. The figure also shows one displacement dial gauge installed to monitor the horizontal displacement for every test configuration.

4.2 Prototype Results

Five configurations were tested using the prototype and that for verification purposes required for the 3D model. In the first configuration the lateral pressure on the wall stem was measured with no EPS buffer (control test), whereas in the other four configurations EPS inclusions with t/h equal to around 0.025, 0.05, 0.10, and 0.13 (density = 20 kg/m^3, to be similar to the EPS used in the 3D model). From Figs. 6a to d show the relationship between the lateral stresses to the wall depth (lateral stresses along the horizontal axis and the wall depth along the vertical axis). As can be seen from the figures, the lateral stresses increased by increasing the excavation level until reaching the maximum designed excavation depth, then the curves started to decreasing at the bottom third of the wall height. By using larger thickness of EPS, a significant reduction in the earth pressure was noticed. This reduction can be clearly seen when comparing between different test configurations presented from Figs. 6a to d. By increasing the EPS thickness to 130 mm, the reduction in lateral stress at a wall depth of 65 cm was 37% and at the wall base was 77%.

4.3 FE Model Verification

From Figs. 6a to d also show a comparison between prototype results and outcomes from the 3D numerical model using EPS thickness of 25, 50, 90, and 130 mm. By comparing the outcomes from the 3D model with the prototype results, it was noticed that there is a good agreement between both; although, some differences appear at the bottom third of the wall height. These differences ranged from 24% to 40%, whereas the FE model results were on the conservative side in most cases except when the EPS

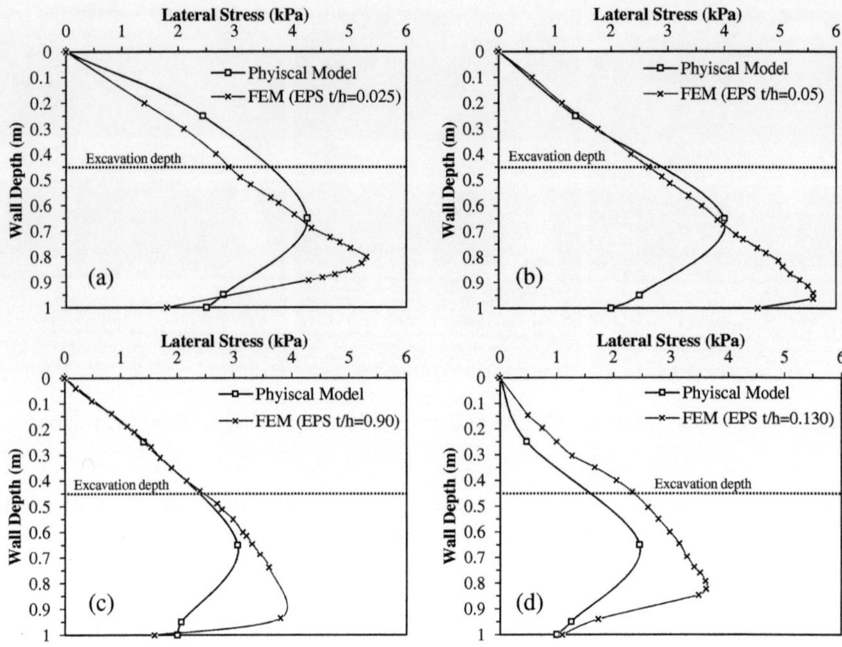

Fig. 6. Comparison between FE and prototype using EPS with various thicknesses

thickness ration t/h exceeded 0.13. However, the profile of the lateral stress acquired from the model was still significantly reduced by using EPS buffers, and the percent of reduction increased by increasing the EPS thickness, following the same trend measured using the physical prototype. Hence, the 3D model showed an acceptable accuracy with the prototype results along the upper two-thirds of the wall and proved that a significant reduction can take place by using EPS buffers behind diaphragm walls.

5   Conclusions

In this study, EPS Geofoam buffers with various thicknesses were introduced behind diaphragm walls to reduce the acting lateral earth pressure. Major findings are presented below.

1. Using an EPS inclusion to act as a buffer between diaphragm walls and soil can reduce the acting lateral earth pressure mainly due to soil arching mechanism.
2. Reduction in the lateral pressure can reach up to 37% by using EPS buffer with thickness 130 mm (i.e., thickness to wall height ratio t/h = 0.13).
3. The numerical 3D model provided relatively conservative but acceptable results compared with the physical prototype outcomes.
4. More research is needed to determine the method of statement for EPS buffer installation during the construction of diaphragm walls.

Acknowledgements. This study is part of the ongoing research project "development of EPS Geofoam to improve infrastructure efficiency", which is funded by the national Science and Technology Development Fund (STDF) – project number 12629. The authors are grateful to the British University in Egypt (BUE), which hosted the tests in its laboratories. Special thanks to Prof. Hani Amin (BUE) for help and fund provided during the prototype instrumentation.

References

AbdelSalam, S.S., Azzam, S.A.: EPS resistance factors and applications on flexible walls. In: ASCE Proceedings, Geo-China 2016: Geosynthetics Civil Infrastructure, Disaster Monitoring, and Environmental Geotechnics, Shandong, China, pp. 131–138 (2016)

AbdelSalam, S.S., Azzam, S.A., Fakhry, B.M.: Reliability and 3D modeling of flexible walls with EPS inclusion. Int. J. Geomech. (2016). ASCE. Published online: 2 December 2016

AbdelSalam, S., Azzam, S., Abdel-Awad, S.A.: EPS Geofoam to reduce lateral earth pressure on rigid walls. In: International Conference on Advances in Structural and Geotechnical Engineering, Hurghada, Egypt (2015)

Athanasopoulos, A.Z., Lamote, K., Athanasopoulos, G.A.: Use of EPS Geofoam compressible inclusions for reducing the earthquake effects on yielding earth retaining structures. Soil Dyn. Earthq. Eng. **41**, 59–71 (2012)

Azzam, S.A., AbdelSalam, S.S.: EPS Geofoam to reduce lateral earth pressure on rigid walls. In: International Conference on Advances in Structural and Geotechnical Engineering, Hurghada, Egypt (2015)

Brinkgreve, R.B.J., Kumarswamy, S., Swolfs, W.M.: Plaxis 3D Reference Manual Anniversary Edition Version 1. Plaxis Bv, Delft (2015). ISBN-13: 978-90-76016-19-2

Ertugrul, O.L., Trandafir, A.C.: Lateral earth pressures on flexible cantilever retaining walls with deformable Geofoam inclusion. Eng. Geol. **158**, 23–33 (2013)

Horvath, J.S.: Expanded polystyrene (EPS) Geofoam: an introduction to material behavior. Geotext. Geomembr. **13**(4), 263–280 (1994)

Karpurapu, R., Bathurst, R.J.: Numerical investigation of controlled yielding of soil-retaining wall structures. Geotext. Geomembr. **11**(2), 115–131 (1992)

Lutenegger, A.J., Ciufetti, M.: Full-scale pilot study to reduce lateral stresses in retaining structures using Geofoam. Final report, Project no. RSCH010-983 Vermont DOT, University of Massachusetts, Amherst, MA, USA (2009)

Padade, H.H., Mandal, J.N.: Expanded polystyrene-based geomaterial with fly ash. Int. J. Geomech. **14**(6) (2014)

Zarnani, S., Bathurst, R.J.: Numerical modelling of EPS seismic buffer shaking table tests. Geotext. Geomembr. **26**(5), 371–383 (2008)

A Case Study of Efficient Solution for Very High Geogrid-Reinforced Retaining Wall

Izzaldin M. Almohd[1(✉)], Dimiter Alexiew[2], and Rami M. El-Sherbiny[3]

[1] Applied Science Private University, Amman, Jordan
i_ayasrah@asu.edu.jo
[2] Huesker Synthetic GmbH, Gescher, Germany
dalexiew@huesker.de
[3] Cairo University, Giza, Egypt
rsherbiny@eng.cu.edu.eg

Abstract. Geogrid-reinforced earth walls are today's state-of-the art for earth retaining and stabilization systems. They have proven efficient and durable with superior resistance to static, resilient and dynamic loads. Worldwide, many codified design procedures and experience are available. Beside their technical and economic advantages, they also have at least 30% lower carbon footprint than conventional walls or stabilizing systems, e.g. concrete retaining walls, coupled with a great advantage in terms of sustainability. Despite the formal design requirements, the systems are capable of adopting specific local conditions as terrain geometry, as well as aesthetic preferences, and are shown to be of high cost benefit compared to other types of structures. In these aspects, the type of the facing is of special interest as a major decision-making factor. Accordingly, a recently executed project in Saudi Arabia is described in the context mentioned above focusing on specific issues like geometry optimization, facing type, adaptation and some rare specific solutions as well as specific design requirement due to the significant wall height and value of adjacent structures. The provided solution is appropriate for different climates, including the harsh, humid and hot environment in the Middle East.

1 Introduction

The process of implementing a specific type of reinforced soil wall or stabilization structures and the subsequent design are based on a number of factors, such as:

- Suitability and feasibility of the system and cost benefit,
- Safety and adequacy of the design,
- Serviceability and performance requirements or limitations, and
- Site requirements or limitations.

Alexiew (2012), Alexiew et al. (2013), and Guler et al. (2011, 2012) described the benefits of GRS systems in terms of stability, flexibility and sustainability. Currently, there are quite few well-acknowledged and used codes as well as methods and tools for GRS wall design (computer programs or software). Limit Equilibrium Analyses are

© Springer International Publishing AG 2018

M. Bouassida and M.A. Meguid (eds.), *Ground Improvement and Earth Structures*,
Sustainable Civil Infrastructures, DOI 10.1007/978-3-319-63889-8_5

still widely and commonly used for the design process, especially where deformations are either estimated to be small or can be predicted using other procedures or software. Deformation predictions shall take into account the construction progress stages (bottom up construction with progressive and cumulative deformation). Numerical, Finite Element (FE) programs, such as Plaxis 2D, is usually used with proper material models and stage configurations. For walls with simple geometries, simple calculations spreadsheets with cumulative stage-wise calculations are also possible (Al Mohd 2010).

For some walls or systems, excessive deformations are predicted (due to foundation formations, wall geometry etc.) or the walls are constructed in the vicinity of existing structures or having utilities within the area of influence, imposing limitations on wall deformations. As such, additional measures shall be made to either minimize deformation or control deformations occurrence in a timely manner. Minimizing deformations is typically achieved by site preparations (within foundations or retained zones) in addition to strict material specifications and compaction requirements and quality assurance. A far more feasible and cost effective procedure is to control when deformations can occur and minimize their effects on surroundings (other structures and utilities) and the wall facing. The reinforcing geogrids and reinforced fill will undertake majority of the deformation during construction (under geostatic loads) prior to attaching or installing the final protective wall facing.

A case study is presented in this paper that involves a 24.4 m high wall constructed in the hot and humid area of Riyadh, KSA, with a robust, durable and well-finished "Muralex®" steel mesh facing.

2   Case Study

The wall described herein is constructed as part of a mega project in Riyadh, KSA. The wall is nearly 670.0 m long with a maximum height of 22.4 m with additional 1.5 m reinforced concrete wall and traffic surcharge on top (Fig. 1). He facing of the wall is inclined at a slope of 1 horizontal to 10 vertical (1 H: 10 V). The fill used for the construction of the wall (reinforced and retained fill) is rock fill. The wall is founded partly on fill and partly on rock, raising concerns on differential movements (horizontal movement and settlement). A number of alternative retaining wall systems were considered. The Muralex® wall type was selected being the most feasible and efficient system due to construction methodology and sequence which minimizes post-construction differential settlements, flexibility, site adaptability, use of site available material, and expedited construction time.

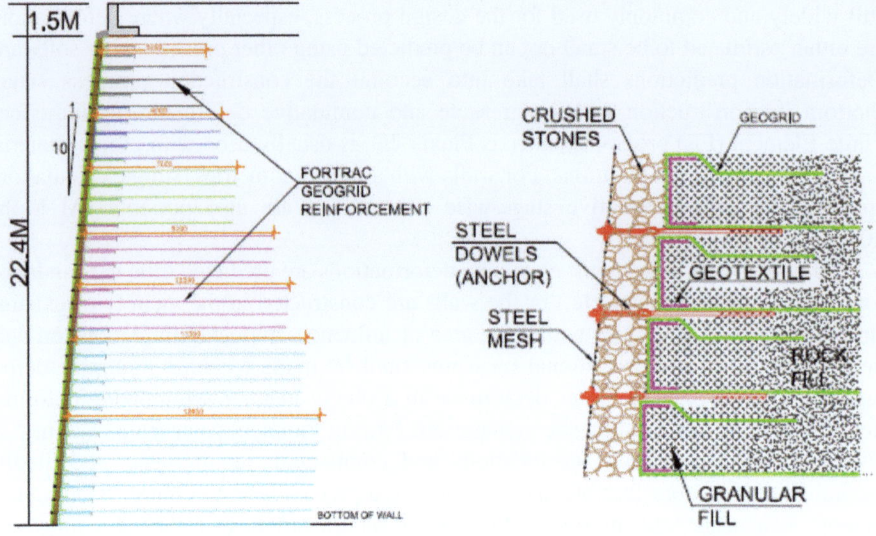

Fig. 1. Riyadh Muralex GRS wall configuration and facing concept

2.1   Description of Muralex GRS Wall

In the Muralex® GRS wall system, the reinforced soil mass is initially constructed with wrap around facing. Steel dowels are installed/embedded into the reinforced soil mass between the geogrid layers for a sufficient length to provide necessary lateral bracing resistance. The dowels will be connected to the final steel mesh at later stages. The general scheme and concept of the Muralex® facing is shown in Fig. 2.

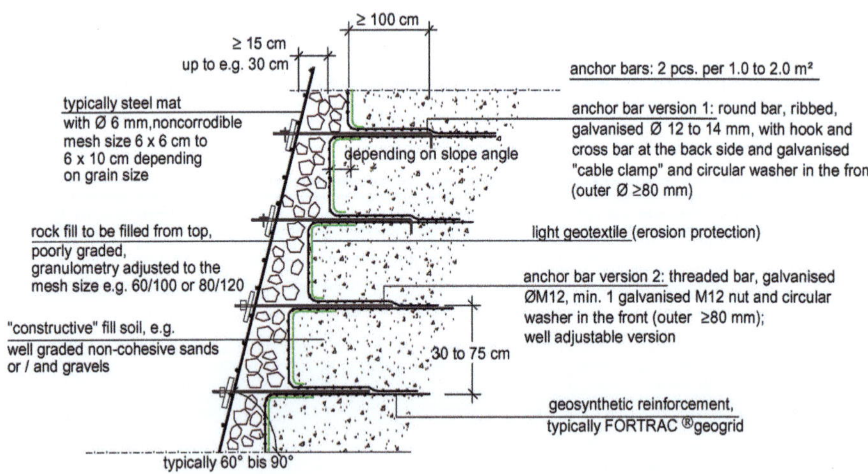

Fig. 2. General concept of Muralex GRS wall and facing

The main advantages of the Muralex system are:

- Geogrids as key reinforcing elements are "hidden", thus protected against UV impact, fire, vandalism and high temperature impact in hot sunny climates;
- Facing can be installed at any point in time following construction e.g. even weeks after completion of the geogrid reinforced package, thus wall deformations occur before facing installation;
- The facing is a "hanging facade" without a bearing function for the GRS, elements can be exchanged at any time, e.g. after a vehicle crash against the facing;
- Flexible and ductile behavior in general, which adapts to total and differential settlements, and seismic loads and impact as well.

2.2 Considerations and Design

Due to project sensitivity, wall height and configuration, foundation soil conditions, and sensitivity of adjacent buildings the wall design carefully considered both stability and deformability criteria. Thus, design calculations were conducted using limit equilibrium (LE) analyses considering allowable stress design (ASD) for overall stability, in addition to numerical simulation using finite element (FE) modeling for performance and deformations evaluations.

2.2.1 Limit Equilibrium Analyses - ASD Design

The LE analyses were initially performed to verify the adequacy of the initially assumed geogrid reinforcement configuration and layout (types, strength, lengths and spacing) considering all design/analyses failure modes (global, internal, compound, sliding, bearing capacity and eccentricity). A customized version of the computer program GGU-international (Huesker stability), was used for design and stability evaluations. This program is capable of analyzing complex geometries with up to 150 layers and types of geogrid reinforcements at variable strength and lengths. The software takes into account both, the stresses and strength of single layers as well as combined reinforcement effects, thus allowing for variable reinforcement lengths. The software also calculates the foundation's bearing capacity, eccentricity, overturning and sliding stability. It utilizes the limit equilibrium analyses for internal stability, external stability and the compound (internal and external) stability. The ASD was used for different sections with different heights, foundation material (fill or bedrock) and boundary conditions. Typical results of the LE analyses for internal and compound stability are presented in Figs. 3 and 4.

2.2.2 Numerical Simulation

In addition to the LE analyses for safety and strength design, numerical modeling using FE-analyses was performed to verify the safety and evaluate deformations of the proposed retaining structure. FE analyses were performed using the computer program Plaxis (v. 8.5) on selected sections that are considered critical either in terms of stability/safety or where deformations are expected to be highest. The FE analyses modeled the planned construction stages to represent the progressive loading nature

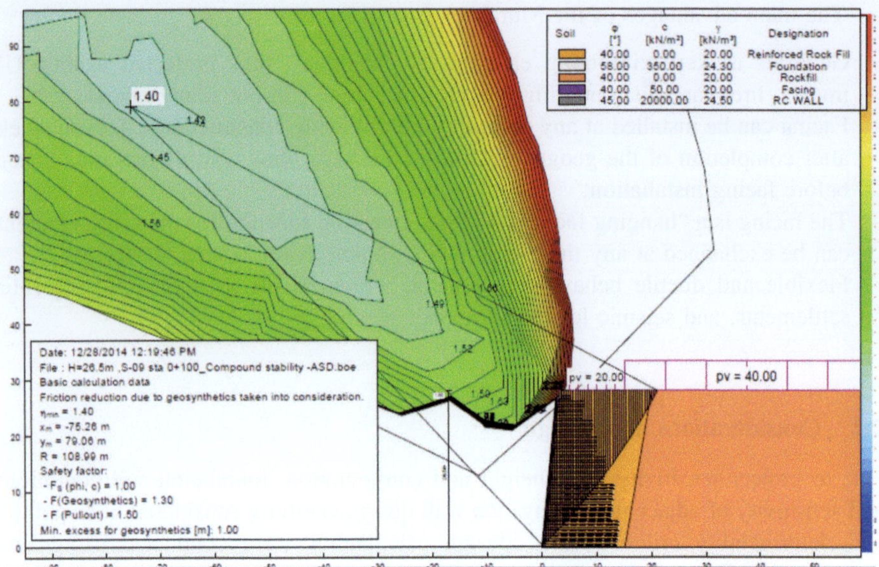

Fig. 3. Typical results of limit equilibrium analyses for compound stability

and accommodate the actual construction steps, which were provided in the construction method statement.

Plaxis includes a linear-elastic perfectly plastic material model for geogrids based on the axial tensile stiffness of the geogrids in unit of kN/m. The steel mesh is modeled as a plate for which the equivalent Young's modulus (E), cross-sectional area (A) and moment

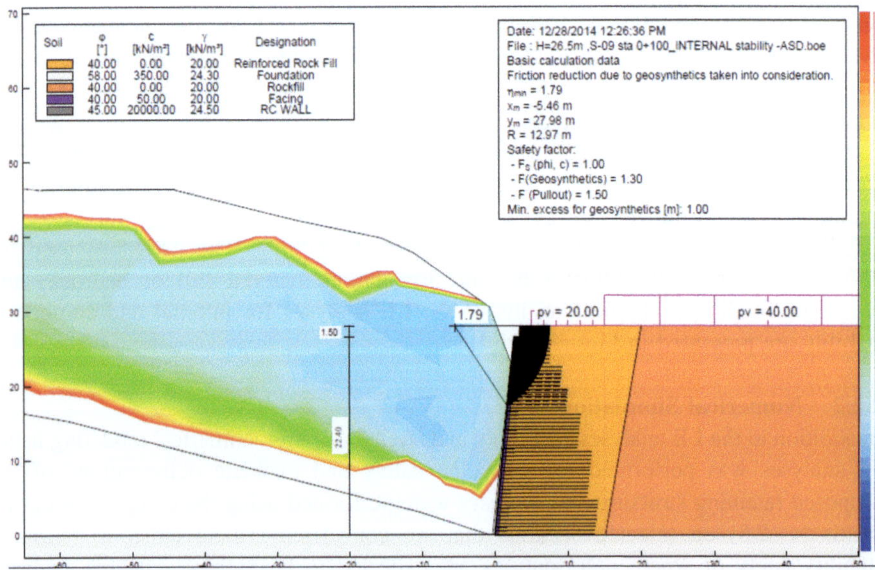

Fig. 4. Typical results of limit equilibrium analyses for internal stability

of inertia (I) are provided. Steel mesh properties were calculated based on the bar diameter and horizontal spacing. The steel anchors (embedded dowels) are also modeled as linear-elastic material. The Hardening Soil (HS) model was selected to simulate the behavior of existing soil strata. The hardening soil model is a non-linear isotropic hardening plasticity model capable of simulating the non-linear behavior of existing soils and accounts for variation of stiffness with effective stress. This modeling approach is capable of capturing behavior of complex walls (El-Sherbiny et al. 2013). The computer program Plaxis evaluates the stability using the friction and cohesion (phi-c) reduction method. Fifteen node triangular elements were used to generate a computer generated randomized mesh with local refinements to optimize calculation time.

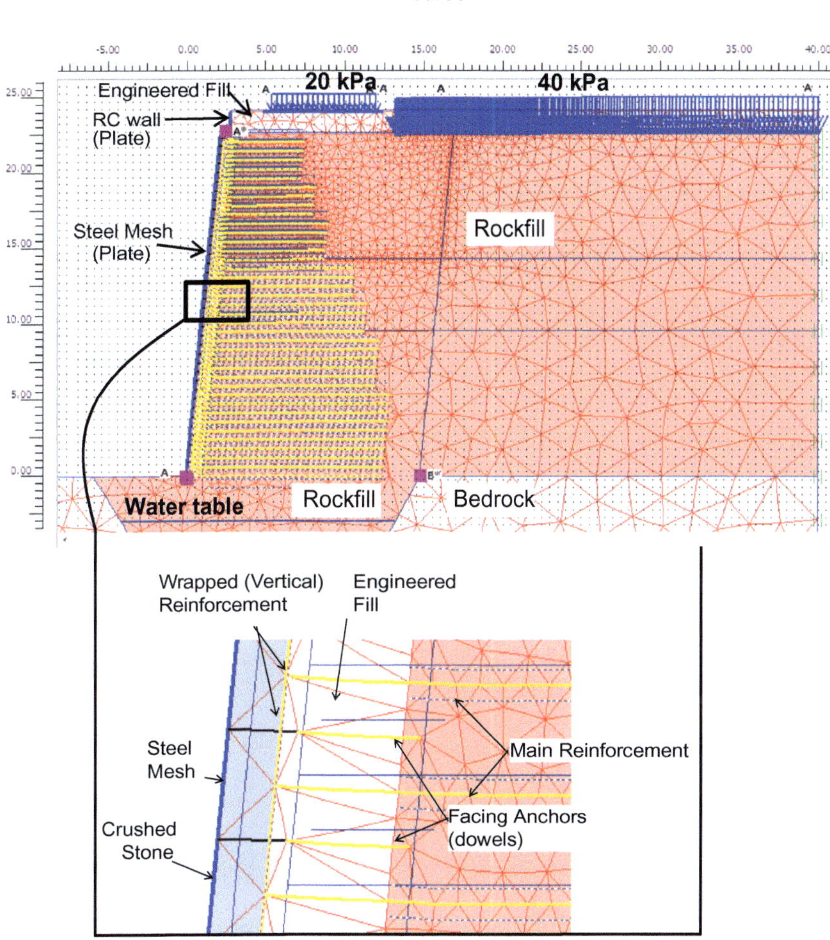

Fig. 5. Plaxis model for Riyadh Metro project

The geometric model and layer description are depicted in Fig. 4. Some of the FE analyses results are also enclosed in Figs. 5, 6 and 7. Based on these results, the maximum horizontal deformations were estimated to be in the range of 11.0 to 12.0 cm, and the factor of safety is 2.0 (Fig. 8). The maximum vertical displacement is about 10.0 cm, which is mainly compression of the thick fill mass Fig. 7.

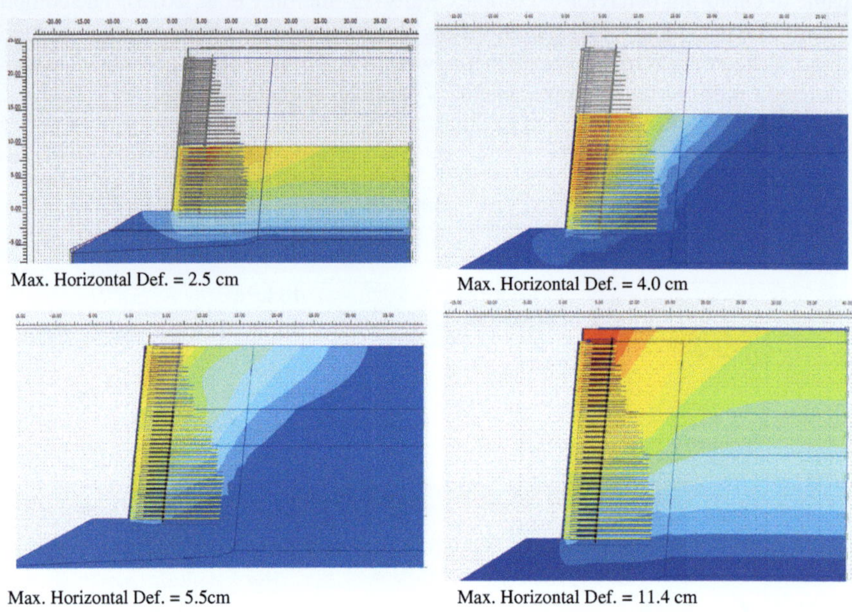

Fig. 6. Maximum horizontal deformations at different stages estimated by Plaxis

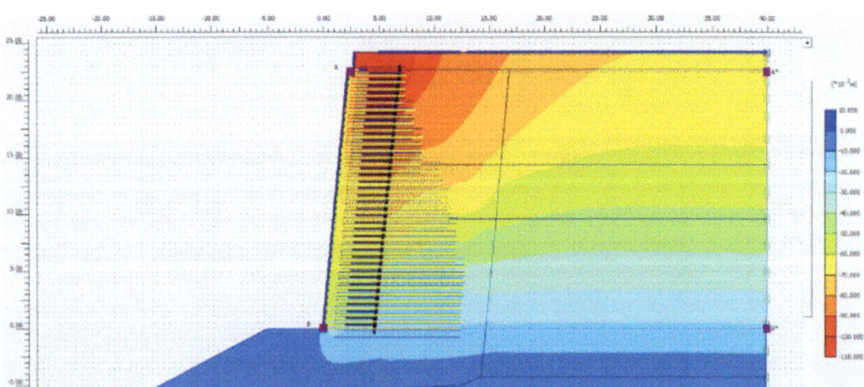

Fig. 7. Vertical deformations contour

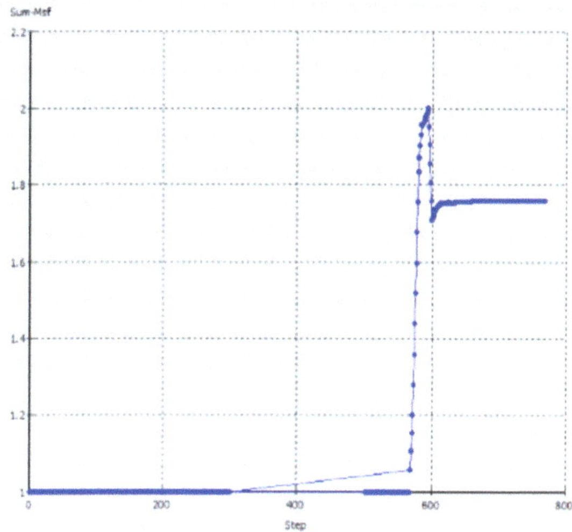

Fig. 8. Calculated factor of safety (Plaxis Model): FS = 2.00

The wall was completed by end of 2015. Survey measurements made during and after completion of the wall revealed horizontal deformations of less than 5.0 cm within the upper parts of the wall, which is less than predicted values. Photographs taken at different stages of construction are depicted in Figs. 9, 10, 11, 12, 13, 14 and 15. In our opinion, they are more useful than any long textual description.

Fig. 9. Installation of geogrid reinforcement layers

Fig. 10. Installation of anchor dowels/bars on top of wrapped back portion prior to installing subsequent reinforcement layer

Fig. 11. Wrap around wall, during construction prior to installation of Muralex facing; anchoring bars visible

Fig. 12. Installation of Muralex steel mesh

Fig. 13. Filling of stone between wrap around GRS and steel mesh to form Muralex facing

Fig. 14. Finished wall, projection-1

Fig. 15. Finished wall with curves and corners

3 Conclusions

A practical case study of an innovative, robust yet flexible, durable, and environmental friendly earth retaining and stabilization system was presented. The wall is of significant height, constructed in high humidity, high temperature area using the Muralex® system. The design process of the wall included numerical simulation for evaluating the deformations and verifying stability and high safety. The measured surveys during construction and after completion showed small deformations, which were lower than calculated from numerical simulations. The wall featured an engineering monument with well-finished surface, and accurate and straight face. This is mainly attributed to the construction methodology of the innovative Muralex® system for which the final facing is installed after wall deformations are mostly completed.

Acknowledgements. The project and this paper would not be successful without the competence, enthusiasm and informal collaboration of designers, consultants, owner and contractor involved which is highly appreciated and acknowledged. The authors would like to show gratitude to the Applied Science Private University in Jordan for the supporting the first author in time needed for preparation of this paper.

References

Al Mohd, I.A.: Non-typical designs in reinforced earth walls. In: 3rd International Conference on Geosynthetics Middle East, Abu Dhabi, United Arab Emirates, November 2010

Alexiew, D.: High geogrid-reinforced walls with a flexible stone filled facing in a mountainous seismic region. In: Proceedings of the 2nd Pan American Geosynthetics Conference & Exhibition, Lima, Peru (2012)

Alexiew, D., Ayasrah, I.M., Abu-Hassan, M.: Geogrid-reinforced walls with almost vertical facings: specific solutions in different environments. In: 6th International Conference on Geosynthetics Middle East, Abu Dhabi, United Arab Emirates (2013)

El-Sherbiny, R., Ibrahim, E., Salem, A.: Stability of back-to-back mechanically stabilized earth walls. In: Geocongress 2013, Stability and Performance of Slopes and Embankments III, San Diego, California, ASCE, pp. 555–565 (2013)

Guler, E., Alexiew, D., Basbug, E.: Dynamic behavior of geogrid reinforced segmental block walls under earthquake loads. In: Proceedings of the 5th International Conference on Earthquake Geotechnical Engineering, Santiago, Chile (2011)

Guler, E., Alexiew, D., Basbug, E.: The behavior of geogrid reinforced segmental block walls under earthquake loads. Topic: soil improvement and reinforcement. In: Proceedings of the 5th European Geosynthetics Congress, Valencia, Spain, vol. 5, pp. 265–269 (2012)

Numerical Simulations of Ground Improvement Using Stone Columns in "Bouregreg Valley"

Noura Nehab[✉], Khadija Baba, Latifa Ouadif, and Lahcen Bahi

L3GIE Department, EMI, Med V Agdal University, Rabat, Morocco
nouranehab.ehtp@gmail.com

Abstract. The soil is generally a heterogeneous material with very variable characteristics. Soil's main issues are reflected through low bearing capacity and significant deformations (settlement) under loads. For Geotechnical experts, such problems present a challenge in many Moroccan regions like Rabat (Bouregreg Valley) and the northern regions. To overcome these defects, stone columns are used as a soil improvement technique to reduce settlement and to increase the foundations' bearing capacity on soft clay soil. The benefits arise from the fact that there is a partial replacement of the compressible soil by a more competent material (compacted stone aggregates). Moreover the stone columns are highly permeable and act as vertical drains facilitating the soft soil's consolidation and improving the foundation's performance. Numerical modeling is a simple and effective method to approach the real behavior of soils reinforced by stone columns. It allows settlement analysis, lateral deformation, vertical and horizontal stresses in order to understand the behavior of columns and soil. It also has the advantage of integrating the settlements of the underlying layers, especially those of least resistance. The aim of this paper is to study the numerical simulation's results. The properties of the soft soil correspond to "Bouregreg valley"-soil. This paper is structured as follows: the first part provides the soil conditions and the parameters related to columns and the second part presents 3D finite element analyses that study the stone columns' performance. These 3D analyses aim to clarify the most important parameters, the influence of the constitutive models and the column geometry. The study also shows the mechanisms of functioning and interactions of stone columns vis-a-vis the various parameters characterizing the granular material "ballast" and the surrounding soil.

1 Introduction

The ground in the "Bouregreg" valley is manifested by weak mechanical characteristics; deformation under significant loads and low bearing capacity; which poses a real challenge for the geotechnical to overcome these defects. Various soil-building techniques have been developed and applied over the last years. Amongst the various ground improvement techniques used, stone columns are the most popular. This is due to their ease of construction, simplicity and economy.

© Springer International Publishing AG 2018
M. Bouassida and M.A. Meguid (eds.), *Ground Improvement and Earth Structures*,
Sustainable Civil Infrastructures, DOI 10.1007/978-3-319-63889-8_6

Stone columns are used as a ground improvement technique which is effective for reducing settlements and for increasing the overall strength, stiffness and the time rate of the soft soil consolidation, accordingly, an acceleration of the settlement. This technique allows also improving the bearing capacity, the slope stability and reducing the liquefaction potential.

They act mainly as inclusions with a higher stiffness, shear strength than the natural soil. Moreover the stone columns are highly permeable and act as vertical drains. Therefore, they facilitate the soft soil's consolidation and improve the foundation's performance (Hughes and Withers 1974; Christoulas et al. 2000; Muir Wood et al. 2000; Guetif et al. 2007; Black et al. 2007).

The majority of analytical methods pertain to primary settlement only (Priebe 1995; Pulko et al. 2011). Current works on modeling, testing, and analysis of compressible soil reinforced with stone columns were reported by McCabe et al. 2009; Mokhtari and Kalantari 2012; Najjar 2013; Balaam and Booker 1981; used the concept of unit cell to study the capacity of single stone column.

Sexton et al. 2013; demonstrate that some recent analytical methods (Castro and Sagaseta 2009; Pulko et al. 2011) are successful. These analytical methods can be implemented mathematically using a unit cell model. However, unit cell models cannot capture the behavior of peripheral group columns beneath footings.

The numerical modeling of a group of stone columns has predominantly been the focus of several studies: Shahu and Reddy 2011; Wehr 1999; Muir Wood et al. 2000; Klai et al. 2015; investigated the performance of groups of stone columns using finite-element analysis.

Killeen 2014; studied small groups of stone columns using 3D finite element analyses. They studied the influence of geometric and material properties. However, there is still lack of information regarding the influence of some geometric and material parameters.

This paper aims to study the numerical simulation results, also, the investigation of the treatment of compressible soil by stone columns technique. The properties of the compressible soil correspond to "Bouregreg valley"-soil. The modeling was done using the 3D finite element method.

To clarify the most important parameters and the influence of the number of columns and their position, 3D numerical analyses were carried out, the study presents the finite element (numerical) models, the parametric study and their results.

We have performed a series of three-dimensional modeling of stone columns with isolated column in the center and then to column groups, using the PLAXIS 3DF software.

2 Site Characterization

A campaign for recognition was implemented at the site of the "Bouregreg valley". Core drilling, with removal intact and reshaped samples (Fig. 1) allowed making laboratory tests (identification and characterization of soils), and establishing a detailed section of land. Pressuremeter tests were realized to determine soil pressuremeter characteristics.

Fig. 1. Soil reinforcement of the Bouregreg Valley and removal of samples for laboratory tests

The soil at "Bouregreg valley" consists of silty clay about 2 m thick, which is underlain by about 8.5 m of soft clay; between 6.5 and 10.5 m, the clay layer has some occasional thin sandy layers. Below a depth of 10.5 m the soil gets stiffer (consistent sand) (Table 1).

Table 1. Characteristics of the "Bouregreg valley" soil

Soil layers	Depth (m)	γ en kN/m^3	E (kPa)	Φ (°)	c (kPa)
Silty clay	2	18	2295	15	25
Clay	6.5	18	2585	15	25
Clay with sandy layers	10.5	19	18334	23	0.5

In the simulations the columns were assumed to reach this rigid soil at 10.5 m depth. The ground water level is at the surface (γ_w = 10 kN/m^3). We are interested in the layers of low geotechnical properties, whose characteristics are determined in Table 1.

We used the following formulas for the calculation of Young's modulus:

$$E = E_{oed} \frac{(1 - 2v)(1 + v)}{1 - v} \qquad (1)$$

With

v: Poisson's ratio

E_{oed}: Œdometric modulus calculated based on the pressuremeter modulus by the relationship:

$$E_{oed} = \frac{E_M}{\alpha} \qquad (2)$$

In which α is a rheological coefficient that depends on the type of soil and the limit pressure.

Stone columns: 10.5 m deep with an average diameter of 0.8 m and a spacing of 1.5 m.

3   Reference Cases

In the first case (Fig. 2); the reference consists of only one stone column under the center of a square rigid footing. The footing width, B, is 1.5 m (B = s) and the column diameter is equal to 0.8 m (dc = 0.8 m). The column is considered to reach a rigid substratum at 10.5 m depth. So, the column is end-bearing (L/H = 1) and has length of L = 10.5 m.

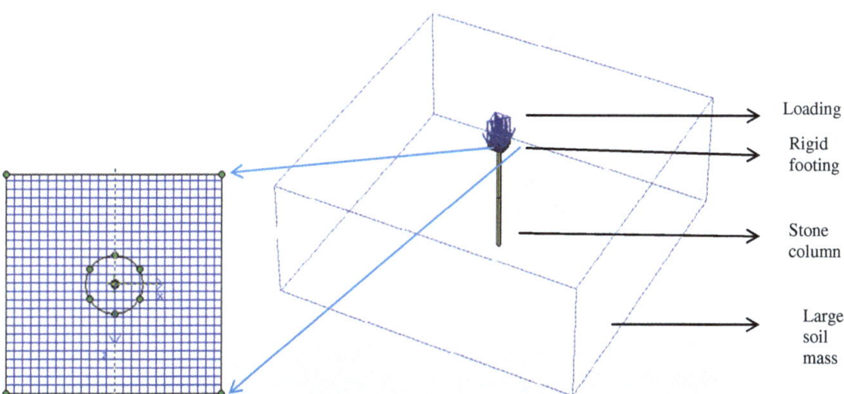

Fig. 2. Numerical model (Plaxis 3DF) of an isolated stone column.

The column is modeled in the center of a large soil mass (36 m $\times$ 36 m $\times$ 10.5 m) (Fig. 2) to avoid the edge effect, the footing of 0.5 m thickness, is considered as a material with properties of concrete: γ = 24 kN/m^3, E = 3.107 kPa and ν = 0.15.

In the second case, we consider a group of columns beneath a rigid square footing (width: B = 4 m). The thickness of the soil layer is taken equal to the length of columns; H = L = 10.5 m (end-bearing columns).

In accordance with standard practice for numerical modeling of stone columns, column-soil interface elements are not used (Shahu and Reddy 2011).

A uniform vertical pressure of P = 100 kPa is applied on the rigid square footing.

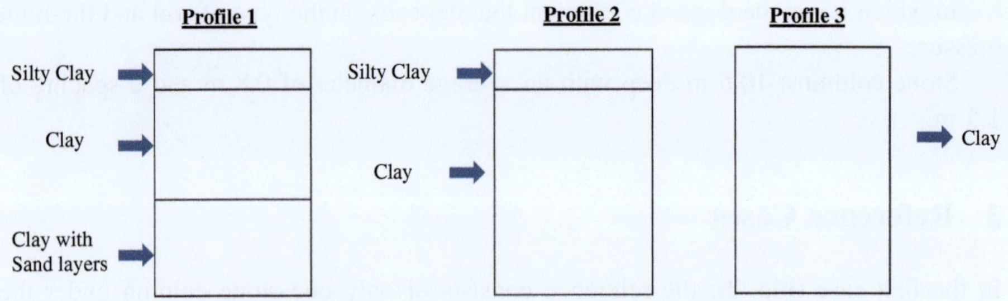

Fig. 3. Various profiles adopted to investigate layers effect

4 Layers Effects

The soil profile adopted for the "Bouregreg valley" test site is modified to investigate the effect of the compressible clay and the other layers (Fig. 3). The settlement's performance for various column arrangements is investigated for the various profiles shown in Fig. 3.

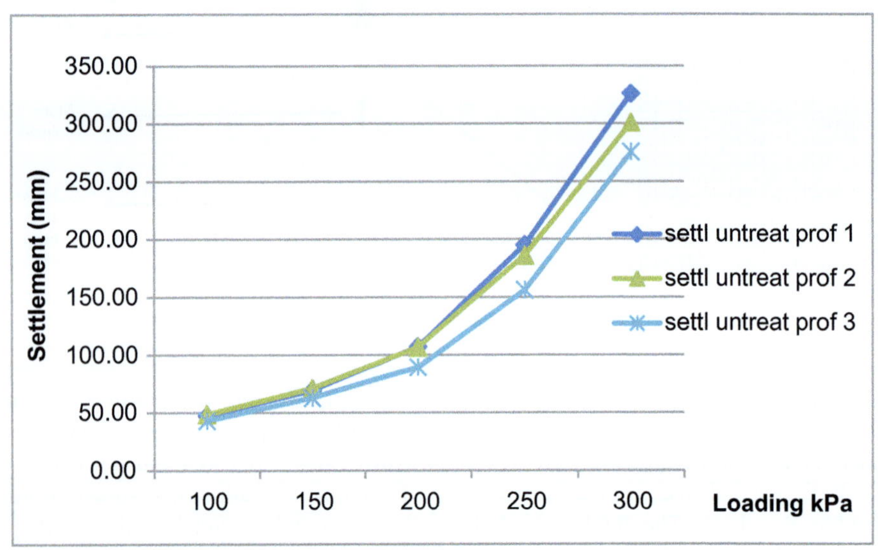

Fig. 4. Settlement of untreated soil for various profiles

The settlement's performance of these profiles is evaluated for an applied pressure varying between 100 kPa and 300 kPa.

Due to soil heterogeneity of the "Bouregreg valley", various profiles were considered. This allows us to study the effect of soil layers.

The graphs (Fig. 4) illustrate for different values of loading, the settlement of three adopted profiles shown in Fig. 3.

The graphs (Fig. 4) show that:

- The settlement of profiles 1 and 2 is higher than profile 3; this is due to the presence of the silty clay layer which is a compressible soil.
- Profile 1 and profile 2 have almost the same graph for a loading less than 250 kPa.
- For small loads settlement is almost the same for all the three profiles, so for reasons of simplification, it may adopt the profiles 2 and profile 3 for small loads.

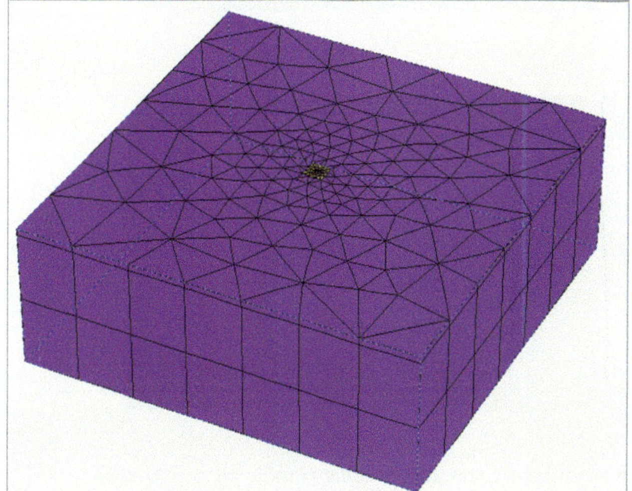

Fig. 5. Typical mesh adopted for finite element simulation, with locally refined mesh in the area of interest (footing and column) (profile 3).

5   Meshing and Boundaries

Plaxis 3DF generate the finite element mesh using a triangulation procedure, but global and local mesh refinements may be defined to ensure a good quality of the mesh, in this way a 3D mesh, composed of 15-noded wedge isoparametric elements is formed, refining the mesh in the area of interest (footing and column) and using a coarse mesh in the far field.

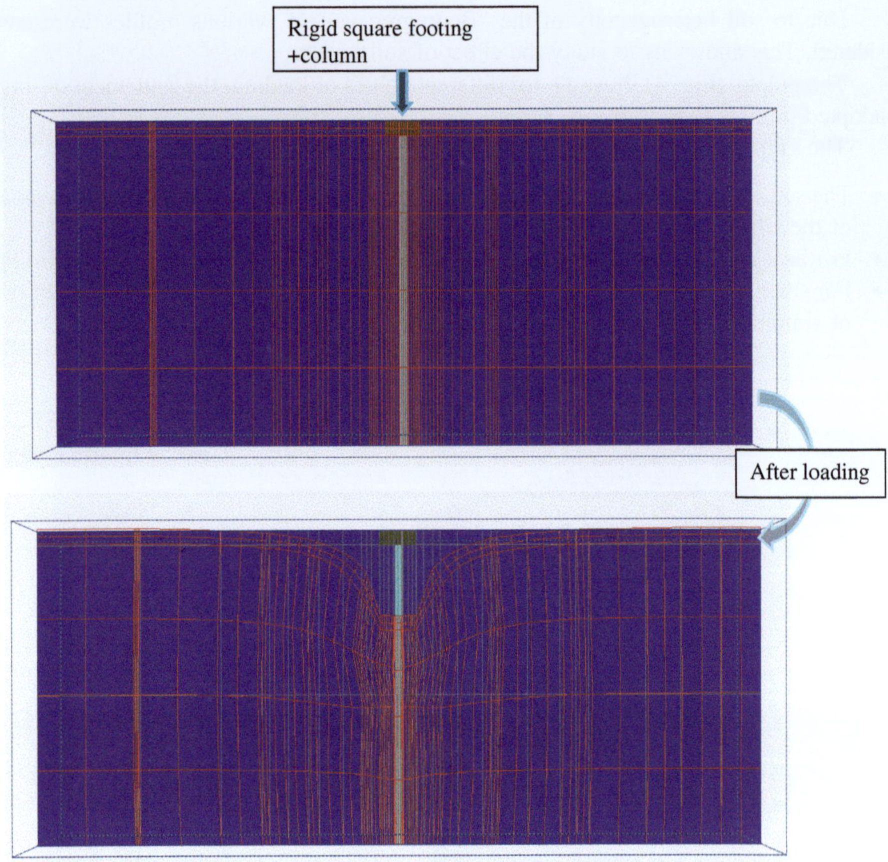

Fig. 6. Deformed mesh before and after loading (Profile 3/P = 100 kPa)

Figure 5 show for the profile 3 the typical mesh adopted for finite element simulation, with locally refined mesh in the area of interest (footing and column).

Adopting profile 3, for an applied pressure equal to 100 kPa: Fig. 6 show the generation of deformed mesh before and after loading.

The lateral boundaries allow vertical movement but restrict radial movement; therefore, it is necessary to position them at a sufficient distance from the footing so that boundaries conditions do not influence the results Killeen (2012). Along the bottom of the soil mass both radial deformation and settlement are restricted (Full fixity is assigned at the base of the geometry).

6 Column–Soil Interface

In keeping with standard practice for modeling stone columns, many authors have declined to use interface elements at the boundary between the granular column material and the in situ soil (Kirsch 2006; Ambily and Gandhi 2007; Domingues et al. 2007; Gäb et al. 2008; Shahu and Reddy 2011), consistent with field observations that during column installation, and due to the column compaction process, the stone columns become tightly interlocked with the surrounding soil due to the lateral displacement (McCabe et al. 2009); No discrete interface zone exists as might be expected for a pile. Perfect adhesion is assumed at the column–soil interface. This assumption of full contact at the column- soil interface is adopted in this study.

7 Constitutive Models

The use of the finite element analysis has become widespread in geotechnical practice an efficient mean of optimizing engineering tasks; however the quality of any prediction depends on the adequate model being adopted in the study.

The choice of a constitutive model depends on many factors but, in general, it is related to the type of analysis that the user intends to perform.

Three sets of calculations were made using different material models on soils: the linear-elastic model, the Mohr-Coulomb model and the Hardening Soil model.

7.1 Mohr Coulomb Model

The Mohr coulomb model idealizes soil as an elastic perfectly plastic material. The behavior of soil before failure is approximated by Hooke's law of elasticity; the failure of soil is based upon the Mohr-coulomb criterion which is defined by two parameters, angle of internal friction and cohesion.

7.2 Hardening Soil Model

The Hardening soil model is an advanced elasto-plastic constitutive model which can be used to simulate the behavior of both the granular column material and the treated soft clay soil; it's an extension of the hyperbolic model developed by Duncan and Chang (1970). The model supersedes the hyperbolic model as it's based on the theory of plasticity rather than elasticity, includes soil dilatancy and introduces a yield cap. It accounts for both shear and volumetric hardening, thus capturing irreversible strains caused by deviatoric and compression loadings, respectively (Schanz et al. 1999).

The hardening soil model was considered appropriate for the granular column material (Sexton et al. 2013). The value of friction angle $\phi = 45°$ has been selected while the dilatancy angle was calculated as $\Psi = \phi - 30°$ (Bolton 1986). A very small value of the cohesion (c = 0.1 kPa) is used to avoid numerical instabilities. A value of m = 0.3 has been used as the power for stress-level dependency of the stiffness. The oedometric modulus $E_{oed}{}^{ref} = 23.3\,\text{MPa}$ was assumed equal to the secant modulus $E_{50}{}^{ref}$ for a reference pressure of $p_{ref} = 100$ kPa, and the unload-reload modulus $E_{ur}{}^{ref}$ was taken as $3E_{50}{}^{ref} = 70\,\text{MPa}$ proposed by Brinkgreve and Broere (2006). A poisson's ratio of $v_{ur} = 0.2$.

For the soft soil: the power is taken equal to m = 1 as recommended by Brinkgreve et al., the reference pressures in the soil and column materials are identical.

8   Comparison of Finite Element Analysis (Plaxis 3DF) with Analytical Methods

The settlement performance of an isolated stone column from Plaxis 3DF is compared with various analytical design methods in this section. The analytical design methods chosen for comparison are:

- Schulze, Baumann and Bauer method.
- Priebe method with assumptions of Ghionna and Jamiolkowski.
- Priebe method.

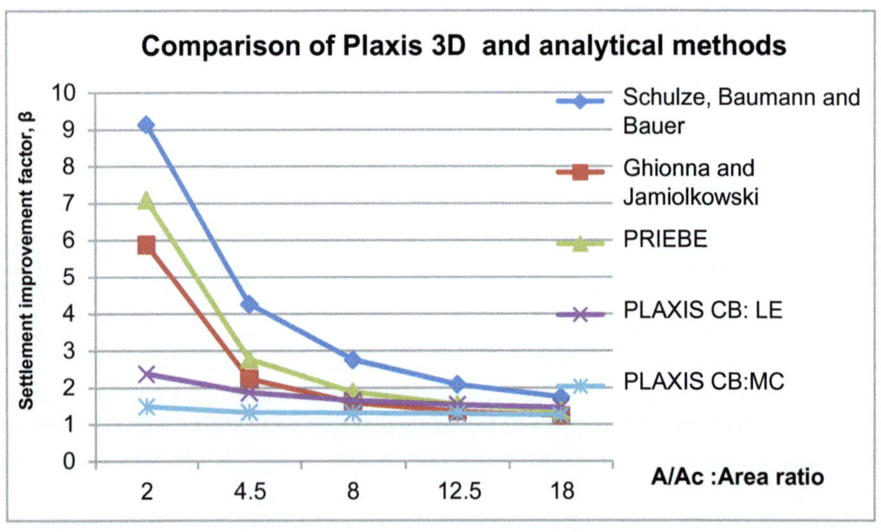

Fig. 7. Comparison of finite element and analytical methods for different area ratio

All of these design methods are based on the unit cell concept, which is consistent with the boundary conditions adopted for the numerical analysis. Analytical methods are based on more simplified models, so we adopt "profile 3" and the case of an isolated column with Mohr coulomb and the linear elastic constitutive models.

The settlement improvement factor β is defined as the ratio:

$$\beta = \frac{\text{The settlement of an untreated soil (no stone columns)}}{\text{The settlement of atreated soil (with stone columns)}} \tag{3}$$

That factor corresponds to the application of a 100 kPa uniform pressure.

The area ratio is defined as the ratio of the footing area (A) to the area of the column beneath the footing (square grid):

$$\frac{A}{A_c} = \frac{B^2}{\pi R_c^2} = \frac{s^2}{\pi R_c^2} \tag{4}$$

With

B: rigid footing width
s: column spacing
R_c: the radius of the column

To compare with analytical methods, the area ratio is varied from 2 to 18 while the width of the rigid footing B (B = s) is varied from 1 m to 3 m. Using both linear elastic and Mohr coulomb rules for stone columns; we obtain variations in the settlement improvement factor, as shown in the graphs (Fig. 7).

The graphs (Fig. 7) illustrate for different values of area ratio the settlement improvement factor, β estimated by different analytical methods:

- Schulze, Baumann and Bauer method
- Priebe method with assumptions of Ghionna and Jamiolkowski
- Priebe method
- 3D finite element.

The examination of the graphs (Fig. 7) shows that:

- The analytical methods tend to over predict the settlement improvement offered by column treatment, especially for small values of area ratio.
- The 3D results using Plaxis 3DF leads to settlement improvement factor values between $1.48 < \beta < 2.25$ for linear elastic column rule, and $1.25 < \beta < 1.50$ for the Mohr coulomb rule.
- The Priebe method and the Priebe method with assumptions of Ghionna and Jamiolkowski gives values close to those obtained by the numerical 3D simulation for $8 < A/Ac < 18$, so these analytical solutions agrees very well with the numerical simulations.

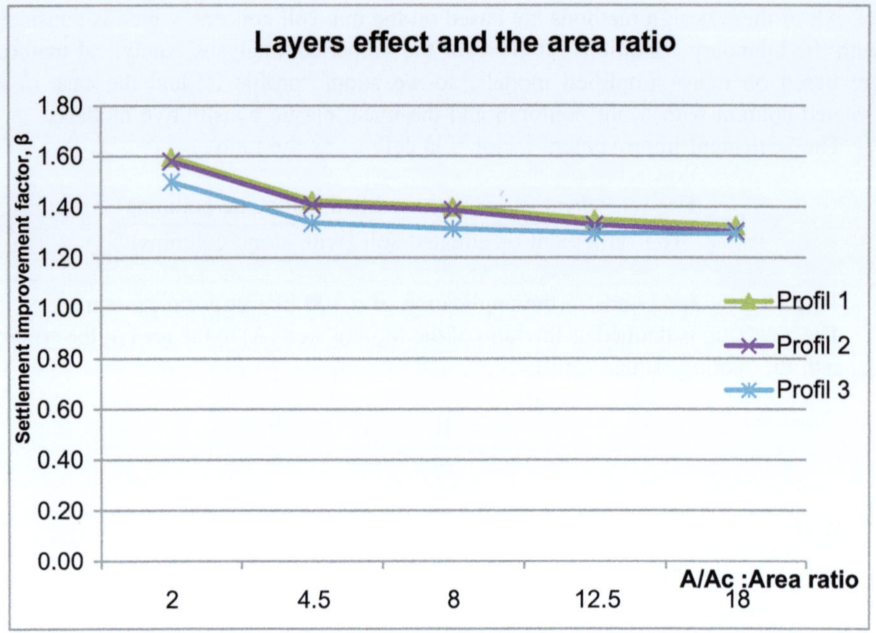

Fig. 8. Settlement improvement for various profiles and area ratio for an isolated column

- Schulze, Baumann and Bauer method over predict the settlement improvement factor (Fig. 7), so we can join the fact that the method of Baumann and Bauer (1974) has a weaker theoretical basis than Priebe (1995) and is believed to give poorer settlement predictions for clayey soils (Slocombe 2001).

9 Parametric Studies (Case of an Isolated Column)

This section analyzes the settlement performance of an isolated column (first reference case), which is determined by the settlement improvement factor, β.

The parametric studies were carried out varying several properties:

  (i)   Layers and the area ratio
  (ii) Friction angle of column material
  (iii) Constitutive models

9.1 Layers and the Area Ratio

This parametric study aims to show the effects of different layers of the soil in the site of "Bouregreg Valley" after treatment by stone columns method. Considering an isolated column (first reference case) and the three adopted profiles: For the stone columns material properties; we adopt the Mohr coulomb constitutive model and an

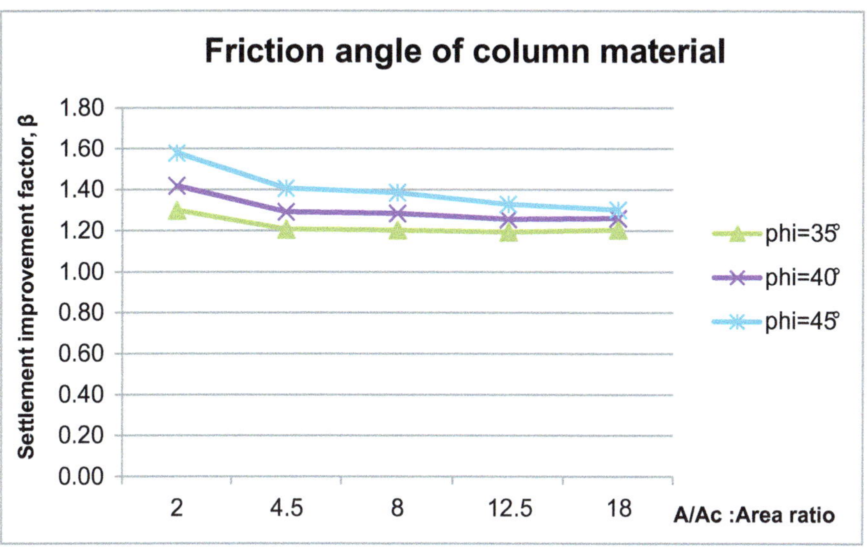

Fig. 9. The influence of internal friction angle of the isolated column and the area ratio

angle of internal friction about 45°. So the settlement improvement factor for an applied pressure of Papp = 100 kPa is determined in Fig. 8.

The "Fig. 8" describes the variation of the settlement improvement factor for the three profiles according to the area ratio.

The examination of the graphs (Fig. 8) leads to observe:

- The settlement improvement factor decreases with the increase of the area ratio for all the three profiles.
- For profile 1: the value of β varies from 1.32 to 1.59, for profile 2: the value of β varies from 1.30 to 1.58, and for profile 3: the value of β varies from 1.29 to 1.50.
- Profile 1 and profile 2 have almost the same graph for 2 < A/Ac < 18.
- So for this case and for the sake of simplicity, it may adopt the profile 2 (for this section).

9.2   Friction Angle of Column Material

To examine the influence of the internal friction angle of column material and the area ratio for an isolated column, we adopt profile 2 and the Mohr coulomb constitutive model for the column material.

The graphs (Fig. 9) illustrate for different values of A/Ac, the settlement improvement factor, β for different values of Φ (35°/40°/45°):

- We observe that the increasing in the internal friction angle of the ballast increases the settlement reduction.

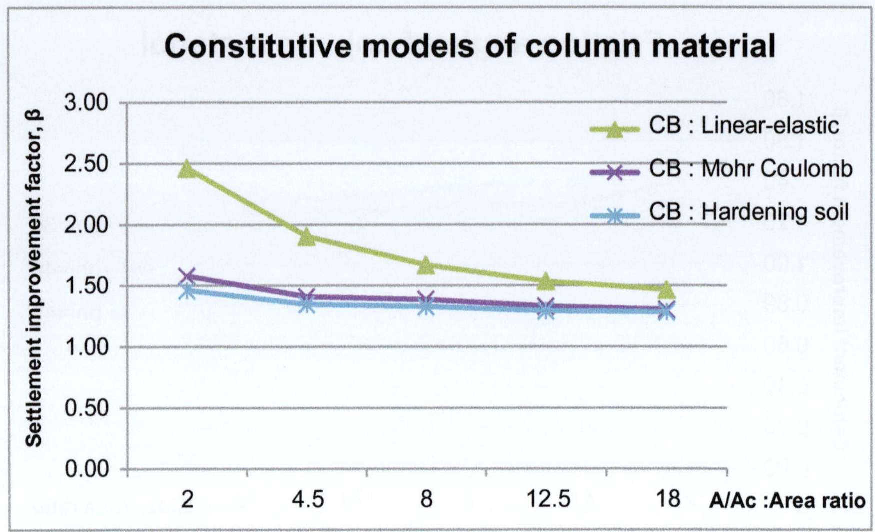

Fig. 10. Settlement improvement for different constitutive models of the column material

- For A/Ac = 2 (low value of area ratio); β varies from 1.30 ($\Phi = 35°$) to 1.58 ($\Phi = 45°$), whereas for A/Ac = 18 (high value of area ratio); β varies from 1.20 ($\Phi = 35°$) to 1.30 ($\Phi = 45°$).
- The settlement increases when A/Ac increases for all values of the internal friction angle of the ballast.
- The graphs approach for high values of A/Ac, so the increasing of ballast friction angle has negligible effects for high values of area ratio.

9.3   Constitutive Models of Column Material

To examine the influence of the constitutive models of column material and the area ratio for an isolated column, we adopt profile 2, the Mohr coulomb model for the soil and $\phi = 45°$ for the column (MC and HS).

The graphs (Fig. 10) illustrate for different values of A/Ac, the settlement improvement factor, β for the behavior rules of column material (Mohr coulomb, Hardening soil and linear elastic models):

- The settlement improvement factor decreases with increasing A/Ac.
- While varying A/Ac from 2 to 18; β varies from 2.46 to 1.46 (for linear elastic model), from 1.46 to 1.28 (for Hardening soil model) and from 1.58 to 1.30 (for Mohr coulomb model).
- The two constitutive models for column material (Mohr coulomb and Hardening soil) reach approximately the same graph.
- The elastic solutions overestimate the settlement reduction, especially that realistic values are used.

10   Parametric Studies (Case of a Group of Columns)

The main focus of this section is to analyze the behavior of stone columns groups in soft soils. A series of finite element analyses using Plaxis 3DF were conducted to examine the influence of several parameters upon the settlement performance.

Using the second reference case (a group of four stone columns), parametric studies were carried out varying several properties:

(a) Column length: L/B, L/dc, L/H (L: the length of the columns, H: soil layer thickness, B: square rigid footing width, dc: column diameter).
(b) Column position (column spacing: s = 1 m/2 m/3 m).
(c) Column compressibility and strength: The friction angle of the column from 35° to 50°.
(d) Constitutive models (linear-elastic, Mohr coulomb and hardening soil behavior rules).
(e) Modular ratio Ec/Es = 15/30/45.
(f) Column confinement.

For a finite group of columns, the area ratio is defined as the ratio of the footing area (A) to the area of the column beneath the footing:

$$\frac{A}{A_c} = \frac{B^2}{N\pi R_c^2} \tag{5}$$

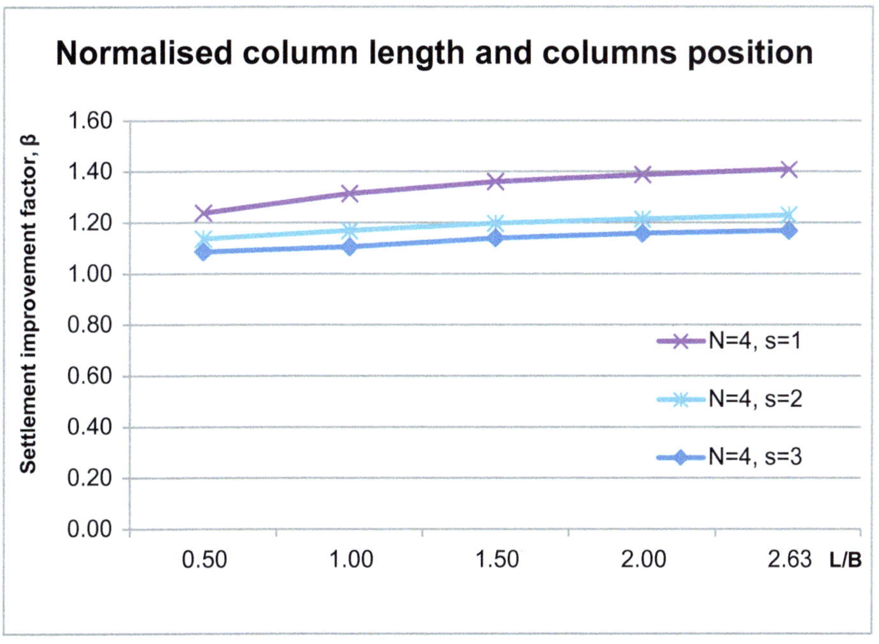

Fig. 11. Settlement reduction for different L/B and columns position

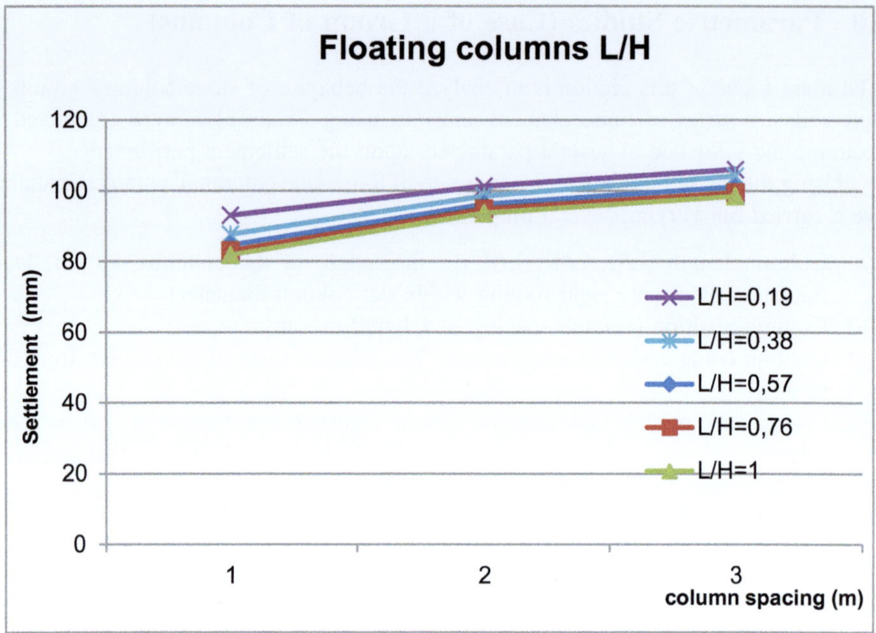

Fig. 12. The effect of floating columns and columns spacing

Table 2. Settlement (mm) for different L/dc and columns position

L/dc	2.27	4.55	6.82	9.09	11.93
N = 4, s = 1 m	93.12	87.74	84.72	83.03	81.85
N = 4, s = 2 m	101.32	98.59	96.22	94.93	93.74
N = 4, s = 3 m	106.06	104.18	101.09	99.52	98.55

With N: number of columns
For the sake of simplicity, we adopt the Profile 3.

10.1   Column Length

Various column lengths were considered which allows studying the effect of L/B, L/H and L/dc, for a group of four columns (Figs. 11, 12 and Table 2):

- For the column numerical simulations, it appears that column spacing (s) and therefore A/Ac governs the variation of settlement improvement factor with column length.
- A hardly noticeable increase in settlement improvement factors is observed from a certain value of (L/B, L/dc) (Fig. 11 and Table 2), that joined the notion of the critical length; as a result, when floating columns reach the critical length, their behavior is similar to end-bearing columns. This value is suggested as 4 to 6 times

Table 3. Settlement (mm) for different friction angle of column material and columns position

ϕ	35	40	45	50
Ψ	5	10	15	20
s = 1 m	94.47	88.18	81.85	75.51
s = 2 m	104.7	98.9	93.74	87.84
s = 3 m	105.73	102.24	98.55	95.22

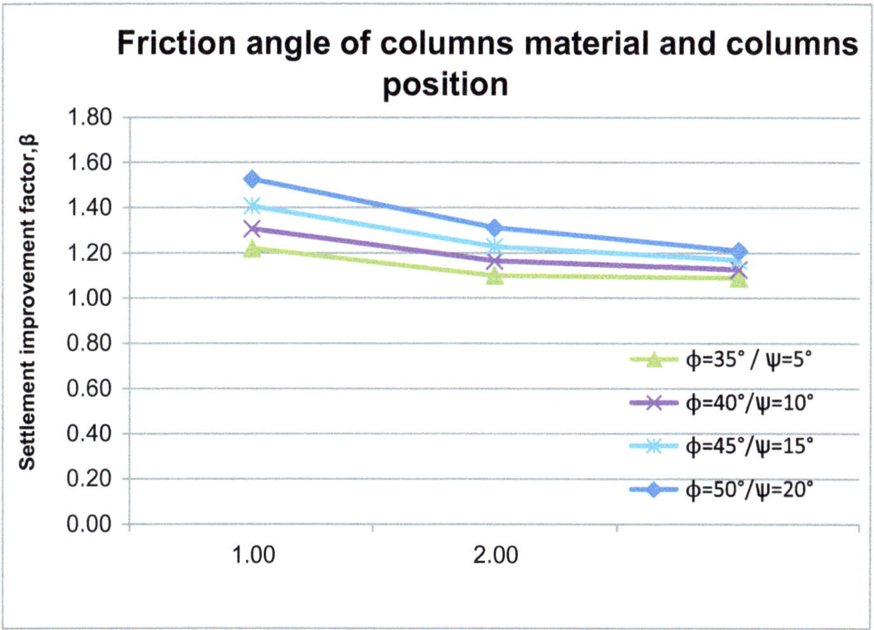

Fig. 13. Settlement improvement for different internal friction angle of columns material and columns spacing

the pile diameter (Hughes and Withers 1974; McKelvey et al. 2004; Black et al. 2007; Najjar et al. 2010).

- It appears that L/B = 1.5 is a significant depth, beyond this level the settlement improvement factor increases very slowly. So L/B = 1.5 is the critical length for the settlement reduction, as proposed by Wehr (2004), it's an acceptable approximation for this case (Fig. 11).
- For floating columns, the column spacing is more relevant than for end bearing columns because the soil layer beneath columns is not improved, so it's able to deform due to the penetration of the columns into the underlying soil (Fig. 12).

10.2   Column Position/Column Compressibility and Strength

This section show the influence of columns spacing, a group of four stone columns is used, Keeping constant A/Ac = 6.57 (η = 15%), so Rc = 0.44 m. The load (Papp = 100 kPa) is applied on the square rigid footing (B = 4 m) and the spacing between the columns is varied: s = 1 m/2 m/3 m.

10.2.1   Friction Angle of Column Material

To examine the effect of the internal friction angle of column material and the columns position for a group of four columns, we adopt profile 3 and the Mohr coulomb constitutive model for both soil and the column material (Table 3 and Fig. 13).

The Table 3 and the graphs (Fig. 13) illustrate for different values of column spacing, the settlement improvement factor, β for different values of Φ (35°/40°/45°/50°):

- We observe that the increasing in the internal friction angle of the ballast increases the settlement reduction: for a columns spacing equal to 2 m; the settlement

Table 4. Settlement (mm) for different constitutive models and column position

	Linear elastic	Mohr coulomb	Hardening soil
s = 1 m	60.71	81.85	239.98
s = 2 m	69.1	93.74	264.53
s = 3 m	85.24	98.55	296.5

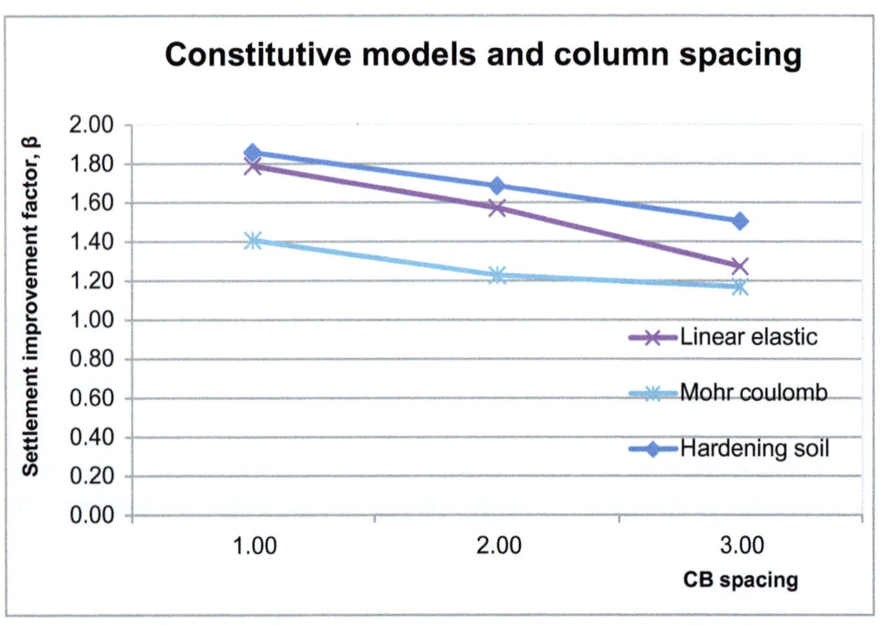

Fig. 14. The influence of constitutive models and columns position

improvement factor (β) varies from 1.10 to 1.31, while varying the internal friction angle Φ from 35° to 50°.

- The settlement increases when the columns are widely spaced for all values of the internal friction angle of the ballast.
- The graphs approach for high column spacing, so the increasing of ballast friction angle has negligible effects for high values of columns spacing.

10.3   Constitutive Models

To determine the influence of constitutive model of both soil and stone column, we consider series of numerical simulations of a group of four columns (profile 3). Adopting Mohr coulomb, linear elastic and hardening soil models (Table 4 and Fig. 14).

The Table 4 and the graphs (Fig. 14) illustrate for different values of column spacing, the settlement improvement factor, β for the behavior rules (Mohr coulomb, linear elastic models and hardening soil).

- The linear elastic model leads to values of β between 1.27 and 1.79 whereas β giving by the Mohr coulomb is between 1.17 and 1.41.
- Numerical analyses adopting the hardening soil model (for both soil and column) give a higher value of the settlement reduction factor than the simulations adopting the elasto-plastic and the linear elastic models, because of the hardening behavior.

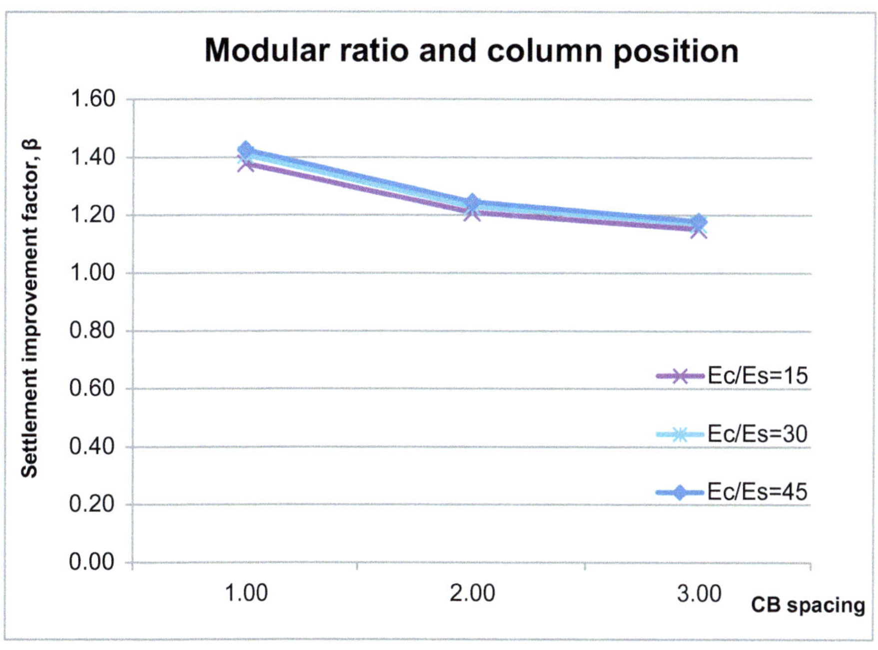

Fig. 15.   The effects of the columns modular ratio and their position

- The settlement improvement factor decreases when the columns are more widely spaced for the three behavior rules, furthermore the both graphs (LE and MC) reach approximately the same value for high columns spacing values.
- For the Mohr Coulomb constitutive law, the settlement improvement factor still very close to the same value between s = 2 and s = 3.

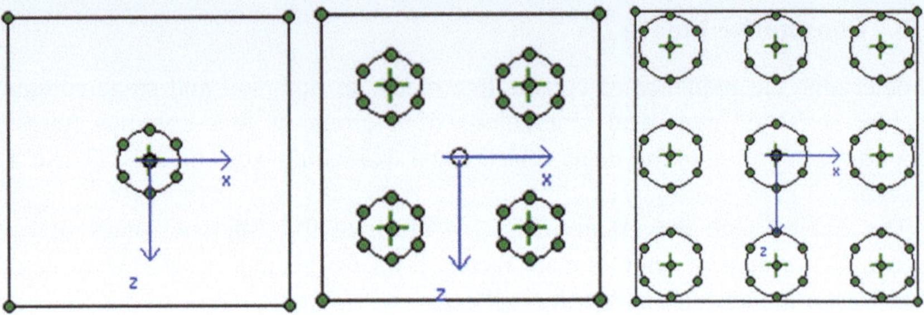

Fig. 16. Column configurations to examine the influence of column confinement

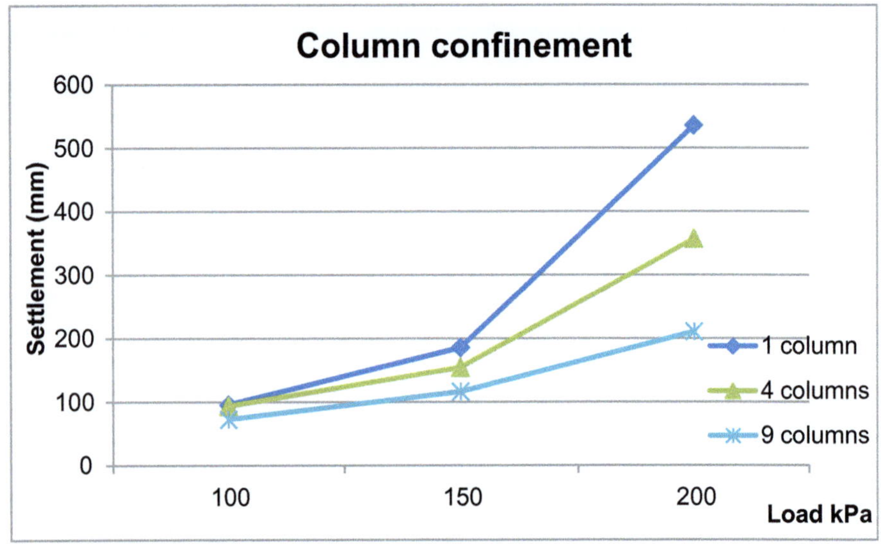

Fig. 17. The influence of column confinement for different applied pressure

10.4 Modular Ratio and Column Position

To determine the effect of modular ratio, various simulations are considered with Ec/Es = 15/30/45, with profile 3 and Mohr coulomb adopted rule (Fig. 15).

The graphs (Fig. 15) illustrate for different values of columns spacing, the settlement improvement factor, β for different values of Ec/Es = 15/30/45.

The results indicate that improvement factors predicted using Plaxis 3DF method increase as the modular ratio increases.

The graphs approach for s = 1 m/2 m/3 m, so the influence of the modular ratio is negligible. Only elastic design methods over predict β values for high modular ratio (Sexton et al. 2013).

10.5 Column Confinement

The confinement depends on A/Ac and the number of columns within a group. The influence of this confinement was determined by analyzing different configurations (Fig. 16)

The number of columns has less influence on the settlement reduction for small values of applied pressure, whereas for high loading the settlement decreases when the number of columns increases (Fig. 17).

11 Conclusions

The stone column behavior was studied using different approaches: Numerical modeling, analytical methods, parametric study concerning the case of an isolated column in the center and then the case of column groups. The comparison of these different methods helps to understand the mechanisms of functioning and interactions of stone columns vis-a-vis the various parameters characterizing the granular material "ballast" and the surrounding soil. The settlement performance of various configurations of columns beneath rigid square footing is examined in this study using the PLAXIS 3DF software.

This detailed analysis of numerical modeling allows The Comparison with analytical methods in the case of an isolated column, which gives early confidence in the ability of Plaxis 3D foundation to capture stone column behavior. Generally the analytical methods tend to over predict the settlement improvement offered by column treatment, especially for small values of area ratio, in this case: Schulze, Baumann and Bauer method over predict the settlement improvement, whereas the Priebe method and the Priebe method with assumptions of Ghionna and Jamiolkowski gives values close to those obtained by the numerical 3D simulation for 8 < A/Ac < 18.

The parametric studies concerning an isolated column were carried out varying several properties: layers, the area ratio, friction angle of column material and the constitutive models; due to the heterogeneity of the soft soil we have adopted three profiles; which allows to determine that the presence of the silty clay layer increase settlements. Properties of column material (internal friction angle, constitutive models) influence the settlement improvement, especially for low values of A/Ac.

For a group of columns, the parametric study were carried out varying several properties: Layers effects, Column position, column confinement and material properties: The friction angle, the modular ratio Ec/Es, the column and the soil behavior rule (linear-elastic, Mohr coulomb and hardening soil models) and the geometric factors: L/B, L/dc, L/H.

Increasing the column material properties (stiffness, friction angle) increases the settlement reduction factor, the influence of these parameters is related to the column arrangement (column position), and to the constitutive models of both columns and soils. Moreover, the column length and the column confinement influence the settlement performance of columns group. A negligible increase in settlement improvement factors is observed from a certain value of L/dc and L/H, and when floating columns reaches the critical length, their behavior is similar to end-bearing columns. This value is suggested by literature as 4 to 6 times the pile diameter. Also, it appears that column spacing (s) and therefore A/Ac governs the variation of settlement improvement factor with column length, furthermore L/B = 1.5 is considered as a significant depth as proposed by Wehr (2004), beyond this level the settlement improvement factor increases so negligently.

The effect of column confinement is also evident as a larger group of stone columns yields smaller settlements.

Acknowledgments. The research was carried out as part of the project of the improvement of soft soils generated by the installation of stone columns under the engineering structures of the "Bouregreg valley". We would like to express sincere appreciation and deep gratitude to all members of 3GIE Laboratory for their efforts and encouragements.

References

Ambily, A.P., Gandhi, S.R.: Behavior of stone columns based on experimental and FEM analysis. J. Geotech. Geoenvironmental Eng. **133**(4), 405–415 (2007)

Balaam, N.P., Booker, J.R.: Analysis of rigid rafts supported by granular piles. Int. J. Numer. Anal. Methods Geomech. **5**(4), 379–403 (1981)

Baumann, V., Bauer, G.E.A.: The performance of foundations on various soils stabilized by the Vibro-Compaction Method. Can. Geotech. J. **11**(4), 509–530 (1974)

Black, J., Sivakumar, V., MeKinley, J.D.: Performance of clay samples reinforced with vertical granular columns. Can. Geotech. J. **44**(1), 89–95 (2007)

Bolton, M.D.: The strength and dilatancy of sands. Géotechnique **36**(1), 65–78 (1986)

Brinkgreve, R.B.J., Broere, W.: PLAXIS 3D Foundation Manual Version 2. PLAXIS BV (2006)

Castro, J., Sagaseta, C.: Consolidation around stone columns. Influence of column deformation. Int. J. Numer. Anal. Methods Geomech. **33**(7), 851–877 (2009)

22 (2000)Domingues, T.S., Borges, J.L., Cardoso, A.S.: Stone columns in embankments on soft soils: analysis of the effects of the gravel deformability. In: Proceedings of the 14th European Conference on Soil Mechanics and Geotechnical Engineering, Madrid, Spain, pp. 1445–1450 (2007)

Duncan, J.M., Chang, C.Y.: Non linear analysis of stress and strain in soil. ASCE J. Soil Mech. Found. Division **96**(5), 1629–1653 (1970)

Gäb, M., Schweiger, H.F., Kamrat-Pietraszewska, D., Karstunen, M.: Numerical analysis of a floating stone column foundation using different constitutive models. In: Proceedings of the 2nd International Workshop on the Geotechnics of Soft Soils – Focus on Ground Improvement, Glasgow, pp. 137–142 (2008)

Guetif, T., Bouassida, M., Debats, J.M.: Improved soft clay characteristics due to stone column installation. Comput. Geotech. **34**, 104–111 (2007)

Hughes, J.M.O., Withers, N.J.: Reinforcing of soft cohesive soils with stone columns. Ground Eng. **7**(3), 42–49 (1974)

Killeen, M.: Numerical modeling of small groups of stone columns. Ph.D. Thesis. National University of Ireland, Galway (2012)

Killeen, M.M., McCabe, B.A.: Settlement performance of pad footings on soft clay supported by stone columns: a numerical study. Soils Found. **54**, 760–776 (2014). http://dx.doi.org/10.1016/j.sandf.2014.06.011

Kirsch, F.: Vibro stone column installation and its effect on the ground improvement. In: Proceedings of the International Conference on Numerical Simulation of Construction Processes in Geotechnical Engineering for Urban Environment, Bochum, pp. 115–124 (2006)

Klai, M., Bouassida, M., Tabchouche, S.: Numerical modelling of Tunis soft clay. Geotech. Eng. J. SEAGS & AGSSEA **64**(4), 87–94 (2015)

McCabe, B.A., Nimmons, G.J., Egan, D.: A review of field performance of stone columns in soft soils. Proc. Inst. Civ. Eng. Geotech. Eng. **162**(6), 323–334 (2009)

McKelvey, D., Sivakumar, V., Bell, A.L., Graham, J.: Modelling vibrated stone columns in soft clay. Proc. Inst. Civ. Eng. Geotech. Eng. **157**(3), 137–149 (2004)

Mokhtari, M., Kalantari, B.: Soft soil stabilization using stone columns-a review. Electronic J. Geotech. Eng. **17**, 1659–1666 (2012)

Muir Wood, D., Hu, W., Nash, D.F.T.: Group effects in stone column foundations: model tests. Geotechnique **50**(6), 689–698 (2000)

Najjar, S.S.: A state-of-the-art review of stone/sand-column reinforced clay systems. Geotech. Geol. Eng. **31**(2), 355–386 (2013)

Najjar, S.S., Sadek, S., Maakaroun, T.: Effect of sand columns on the undrained load response of soft clays. J. Geotechn. Geo-environmental Eng. **136**(9), 1263–1277 (2010)

Priebe, H.J.: The design of vibro replacement. Ground Eng. **28**(10), 31–37 (1995)

Pulko, B., Majes, B., Logar, J.: Geosynthetic-encased stone columns: analytical calculation model. Geotext. Geomembr. **29**(1), 29–39 (2011)

Schanz, T., Vermeer, P.A., Bonnier, P.G.: The hardening soil model: formulation and verification. In: Beyond 2000 in Computational Geotechnics. Ten Years of PLAXIS, p. 281e 90. A.A. Balkema, Amsterdam, The Netherlands (1999)

Sexton, B.G., McCabe, B.A., Castro, J.: Appraising stone column settlement prediction methods using finite element analyses. Acta Geotech. (2013). doi:10.1007/s11440-013-0260-5

Shahu, J.T., Reddy, Y.R.: Clayey soil reinforced with stone column group: model tests and analyses. J. Geotech. Geoenviron. Eng., 1265–1274 (2011). doi:10.1061/(ASCE)GT.1943-5606.0000552

Slocombe, B.C.: Deep compaction of problematic soils. In: Problematic Soils, pp. 163–181. Thomas Telford, London (2001)

Wehr, J.: Stone columns–single columns and group behavior. In: Proceedings of the 5th International Conference on Ground Improvement Techniques, Malaysia, pp. 329–340 (2004)

Wehr, W.: Schottersäulen – das Verhalten von einzelnen Säulen und Säulengruppen. Geotechnik **22**(1), 40–47 (1999). (in German)

Behavior of Pipelines Embedded in Self-compacting Materials Under Traffic Loads

Khalid Abdel-Rahman[(✉)], Tim Gerlach, and Martin Achmus

Institute for Geotechnical Engineering,
Leibniz University of Hannover, Hannover, Germany
`khalid@igth.uni-hannover.de`

Abstract. Self-compacting filling material or controlled low strength materials "CLSM" is a cementitious material which is liquid during filling and is used primarily as backfill e.g. in trenches. Selection of materials should be based on availability, cost, specific application and the necessary characteristics of the mixture, including flowability, strength and excavatability. The main objective of this study is to investigate numerically the behavior of pipelines embedded in self-compaction materials "CLSM" under traffic loading. A two-dimensional numerical model using the finite element system ABAQUS was developed. In this model the material behavior of CLSM is described using an elasto-plastic constitutive model with Mohr-Coulomb failure criterion. The numerical model allows the calculation of bending moment in the pipelines taking the effect of hardening process of "CLSM" on the overall behavior of the pipe-soil-system into account. A subroutine was implemented in ABAQUS in order to quantify the effect of shrinkage of the self-compacting material (CLSM) on the stresses imposed on the pipeline. Finally hints and recommendations regarding the usage of these materials will be given.

1 Introduction

Controlled low-strength material (CLSM) is a self-compacting cementitious material used primarily as a backfill as an alternative to compacted fill. CLSM should not be considered as a type of low-strength concrete, but rather a self-compacting material with features similar to soils. Because CLSM needs no compaction and can be designed to be liquid during filling, it is ideal for use in narrow trenches or restricted-access areas where placing and compacting usual fill is difficult. Conventional CLSM mixtures usually consist of water, cement, fly ash or other similar products and fine or coarse aggregates or both. The investigation presented in this paper deals mainly with the behavior of rigid and flexible pipes embedded in CLSM, taking the effect of hardening and shrinkage of CLSM into account.

© Springer International Publishing AG 2018
M. Bouassida and M.A. Meguid (eds.), *Ground Improvement and Earth Structures*,
Sustainable Civil Infrastructures, DOI 10.1007/978-3-319-63889-8_7

2 Overview of Previous Investigations

Although several studies were conducted on the analysis of buried pipes using soil-pipe interaction theories (e.g. McGrath 1998; Suleiman et al. 2002) a limited number of investigations have explored the performance of pipes buried in controlled low-strength material (CLSM) by numerical simulation (Arsic 2009; Bellaver 2013; Dezfooli et al. 2015).

McGrath (1998) conducted a study on the pipe-soil interaction during backfilling. Diverse backfilling materials were used at varying compaction levels. Several soil box tests and field tests were conducted on steel, concrete, and plastic pipes to compare the results for the different backfilling materials, trench conditions, trench widths, and bedding materials. Suleiman et al. (2002) investigated the effects of large deflection behavior on buried plastic pipes. This study compared the small deflection analysis theory results by using Culvert Analysis and Design (CANDE) software with the large deflection analysis theory.

Different CLSM materials were investigated experimentally and numerically by Arsic (2009). He investigated standard filling materials and CLSM materials in full scale experimental model. Then a numerical model using ABAQUS was developed to predict the behavior of pipelines embedded in CLSM for different trench dimensions. Based on his results the pipeline regulations were adopted and modified for CLSM. Bellaver (2013) developed a three dimensional (3D) nonlinear finite element model of steel pipe coupled with CLSM and compacted soil. The finite element model consists of the pipe and soil interaction during the staged construction of embedment and backfill. The numerical model can be used to predict pipe performance under varying backfill and loading conditions.

Staged construction modeling of steel pipes buried in CLSM was investigated by Dezfooli et al. (2015). The results of field tests were compared with 3D nonlinear numerical model. The comparison of the results indicates that the finite-element model is capable of predicting the deflection of the buried steel pipes in different backfilling and trench configurations. Summarizing, only a limited number of investigations exists. In none of these studies, the effect of hardening process and shrinkage of self-compacting materials in the numerical modeling was considered.

3 Properties of Self-compacting Materials

The properties of CLSM cross the boundaries between soils and concrete. CLSM is manufactured from materials similar to those used to produce concrete, and is placed from equipment in a fashion similar to that of concrete. In-service CLSM, however, exhibits characteristic properties of soils. The properties of CLSM are affected by the constituents of the mixture and the proportions of the ingredients in the mixture. In the following the main features will be explained briefly.

3.1 Hardening Process

One of the most important aspects for CLSM is the hardening process which describes the increase of material stiffness with time. Figure 1 shows the development of Young's modulus (E-modulus) of typical CLSM with time. The E-modulus increases gradually from almost 0.50 MPa up to 120 MPa. Simultaneously the Poisson's ratio (v) decreases from 0.48 (CLSM behaves like a liquid) to 0.20, which corresponds to the solid state like concrete or soil. These material parameters will be adopted for the numerical modeling.

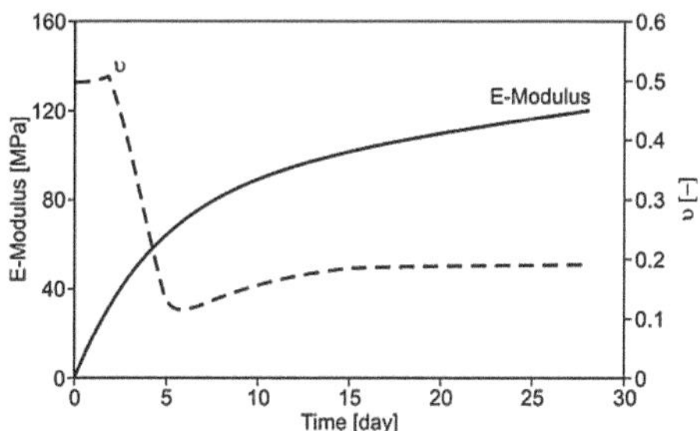

Fig. 1. Dependency of E-Modulus and Poisson's ratio (v) on time (Arsic 2009)

3.2 Shear Strength

Since engineering applications of CLSM as a substitute to conventional compacted fill is growing, it is getting more important to measure CLSM properties in terms of geotechnical engineering parameters by either direct measurement or by developing correlations between geotechnical and concrete test results. The shear properties of CLSM are especially important and can be evaluated using both a direct shear test and a triaxial shear test. The CLSM considered here showed after 28 days an internal friction angle ranging from 30° to 40° and the cohesion ranges from 60.0 kN/m^2 up to 90.0 kN/m^2 (Arsic 2009).

3.3 Shrinkage Process

Compared to concrete, CLSM typically has a very high water-cement ratio and water content, factors that may cause drying shrinkage. Based on experimental investigations (Buhr 2015), the shrinkage of CLSM was about 0.1% till 0.3%. Hardened CLSM may exhibit shrinkage cracks. However, they do not affect the structural integrity of the material for most applications and were not considered in the numerical modeling.

4   Finite Element Modeling

For the investigation of the behavior of pipelines embedded in self-compacting materials, a two-dimensional (2D) numerical model was developed. For the numerical analysis the finite element program ABAQUS (Simulia 2015) has been used.

4.1   Numerical Model

To investigate the deformation response of a pipe with a diameter of 0.30 m embedded in self-compacting material a two-dimensional (plane strain) finite element model was developed. The dimensions of the numerical model used in the analysis are shown in Fig. 2. The elements used to model the soil and CLSM are 6-noded and 8-noded plane strain (CPE6 & CPE8) elements, while beam elements (B22) have been used to simulate the pipelines. In this study, the embedment depth (h) was set 1.0 m and the trench width (b) was 0.90 m. The overall dimension of the numerical model was set to 9.30 m * 3.20 m. With these model dimensions the calculated behaviour of the pipe is not influenced by the boundary conditions.

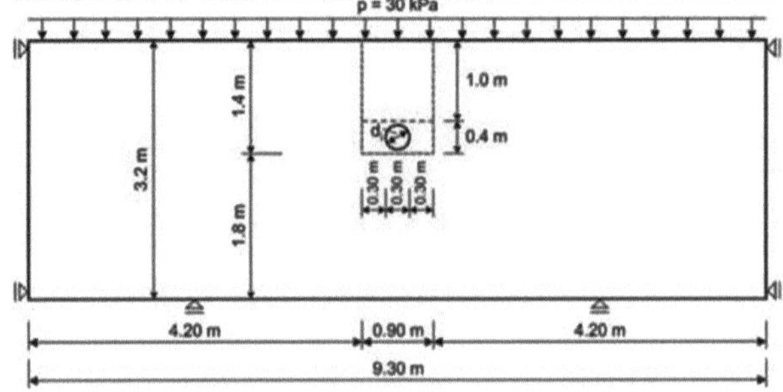

Fig. 2.   Dimensions of the numerical model

4.2   Elasto-plastic Constitutive Model

The most important issue in geotechnical numerical modeling is the simulation of the soil's stress-strain behavior. An elasto-plastic material law with Mohr-Coulomb failure criterion and non-associative flow rule was used to describe the behavior of self-compacting material (CLSM) and also the surrounding sandy soil. For the planned investigation, Mohr-Coulomb constitutive model is sufficient as one of the important aspects in this investigation is the hardening process of self-compacting material and also the modelling procedure is monotonic without any unloading or reloading stages.

According to the pipeline regulations, the pipe trench should be divided into four parts (see Fig. 3), for each part different soil properties should be applied. For the parts E_3 & E_4, sandy soil is assumed. Table 1 summarizes the different parameters required for these two parts (E_3 & E_4). For the other parts inside the trench (E_1 & E_2), self-compacting material (CLSM) will be used.

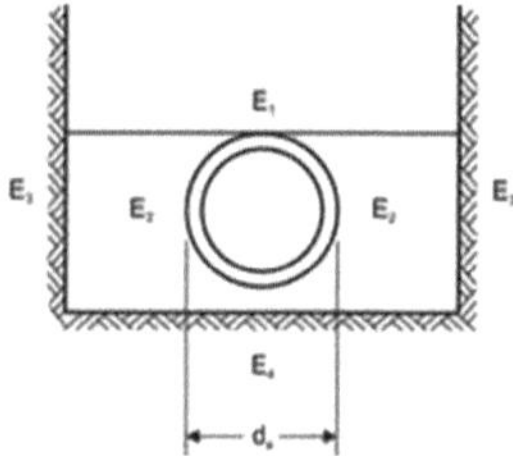

Fig. 3. Pipeline and the different trench parts (E_1, E_2, E_3 and E_4)

Table 1. Material properties for the sandy material outside the trench

Zone	γ (kN/m³)	E (MPa)	v (-)	c' (kN/m²)	φ' (°)	ψ (°)
E_3	18.0	50.0	0.25	1.0	35.0	5.0
E_4	18.0	80.0	0.25	1.0	35.0	5.0

Regarding controlled low strength material (CLSM) for E_1 & E_2, the time dependency for both E-modulus and Poisson's ratio will be implemented as shown in Fig. 1, by tabulating this dependency of the elastic parameters as step-dependent which allows the consideration of hardening process on the behavior of CLSM in the numerical model. The other material parameters required for Mohr-Coulomb material model (internal friction angle φ', dilatation angle ψ and cohesion c') remain unchanged with the hardening process as listed in Table 2.

Table 2. Material properties for self-compacting material (CLSM)

Unit weight γ	18.0 (kN/m³)
Internal friction angle φ'	35.0°
Dilatation angle ψ	5.0°
Cohesion c'	82.0 (kN/m²)

For comparison reasons, two different pipe materials will be modeled. The first one is a rigid pipe made of concrete and the second one is a flexible PVC pipe. Table 3 summarizes the main material properties adopted for the numerical modeling.

Table 3. Material properties for different pipeline materials

Material	γ (kN/m³)	E (MPa)	G (MPa)	v (-)	D (m)	Thickness (m)
Concrete	24.0	31000	12900	0.20	0.3	0.04
PVC-U	14.0	2250	833	0.35	0.3	0.01

4.3   Contact Conditions

To describe the behavior of the embedded pipes accurately two different contact pairs are adopted in the numerical model. The slave-master concept was used, which means that the master nodes can penetrate in the slave and vice-versa is not allowed. The first one (CP1) describes the contact behavior between the pipe and the surrounding soil. The second contact pair (CP2) is implemented to simulate the vertical sliding between the excavation part (trench) and the surrounding soil on both sides of the trench (see Fig. 4). For both of the contact pairs an elasto-plastic model was implemented. The maximum frictional shear stress is dependent on the normal stress and the friction coefficient, which is assumed to $\mu = 0.431$ ($\mu = tan\ (2/3\varphi)$) in case of the first contact pair (CP1) and $\mu = 0.70$ ($\mu = tan\ (\varphi)$) for the second contact pair (CP2).

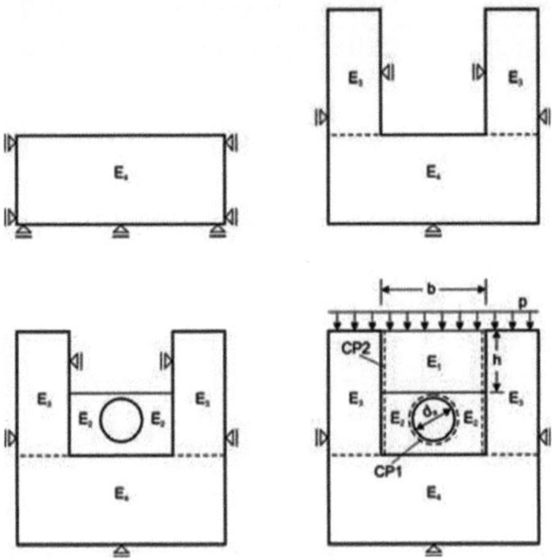

Fig. 4. Numerical modelling steps

4.4   Numerical Modeling Procedure

The modelling process is executed stepwise. Firstly, the primary stress state using own weight of the lower soil medium located beneath the pipeline (E_4) is generated. In the subsequent steps different soil layers are activated to simulate the construction steps as shown in Fig. 4. In the next step, both sides of the trench (E_3) will be activated with the corresponding boundary conditions on the vertical sides. Consequently, the pipeline embedded in the middle part (E_2) and the contact pairs (CP1 & CP2) will be added. Finally, the upper part (E_1) in the trench and the vertical surcharge ($p = 30$ kN/m^2) are applied on the top surface of the model. During the modelling steps, the hardening process which describes the dependency of Elastic modulus and Poisson's ratio (see Fig. 1) is taken into consideration by changing the material properties set for each construction step. Finally, the shrinkage of CLSM was applied by decreasing the dimensions of the CLSM-elements by 0.1% in both horizontal and vertical directions to find out the effect of possible shrinkage on the overall behavior of pipelines.

5   Numerical Modeling Results

The numerical modeling deals mainly with the behavior of rigid and flexible pipes embedded in self-compacting material "CLSM" taking the effect of hardening/shrinkage process into consideration. The vertical stress, horizontal stress and bending moments in the pipelines will be presented and evaluated.

5.1   Concrete (Rigid) Pipe

Figures 5 and 6 show the distribution of contact stress (normal and shear stress) between the pipe and the surrounding soil "CLSM" projected vertically to give the vertical stress component and horizontally to give the horizontal stress component acting on the pipe.

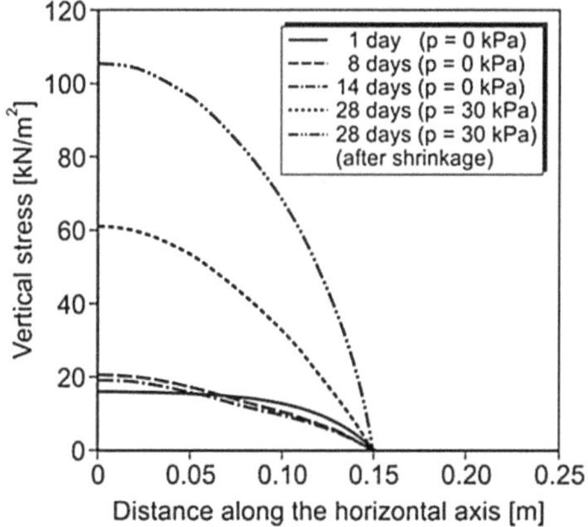

Fig. 5. Vertical stress around rigid pipeline embedded in CLSM

The vertical stress distribution (see Fig. 5) has its maximum value on the vertical (symmetry axis) of the pipe and the minimum value at the end of pipe (x = radius of the pipe = 0.15 m). In the first two weeks, there is a redistribution of the vertical stress and it decreases slightly with the hardening process due the change in the material properties. By applying the traffic load of 30.0 kN/m^2, the vertical stress reaches a maximum value of 61.0 kN/m^2 before applying the shrinkage rate of 0.1% and reaches after the shrinkage 102.5 kN/m^2. These results show that the shrinkage leads to an increase of almost of 65%. This can be explained due to the movement of soil towards the pipe, which consequently induces an increase in the vertical stress on the pipe.

Figure 6 shows the horizontal stress component projected along the diameter of the pipeline (D = 0.30 m). At the crown and the bottom of the pipe, the horizontal stress component is equal to zero. Then the horizontal stress increases gradually to reach its maximum value at almost an angle of 45.0° measured from the vertical axis. As before, in the first two weeks there is redistribution of the horizontal stress acting on the pipe. Then by applying the traffic loading, the horizontal stress reaches 21.0 kN/m^2 before applying the shrinkage and then 54.0 kN/m^2 after the shrinkage (0.1%) was modelled.

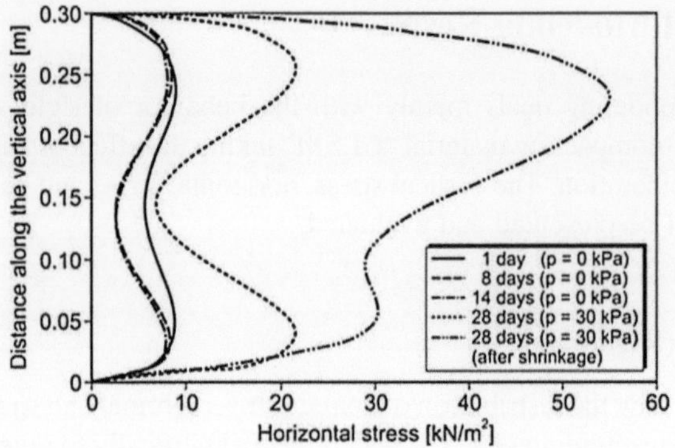

Fig. 6. Horizontal stress around rigid pipeline embedded in CLSM

Figure 7 shows the distribution of bending moment around the pipeline. The x-axis shows the angle measure from the crown (zero at the pipe top, 90° at horizontal pipeline axis and 180° at the bottom of the pipeline). In the first days up to 14 days, the bending moment is comparably small (ca. 0.05 kNm/m) and it decreases accordingly with the hardening of CLSM in the trench. Under the application of the surcharge (p = 30.0 kN/m^2) the bending moment increases up to a maximum value of 0.21 kNm/m. Under shrinkage process, the bending moment increases up to 0.30 kNm/m which is almost 40% higher than the previous value. This means that the effect of shrinkage of CLSM should be taken into account.

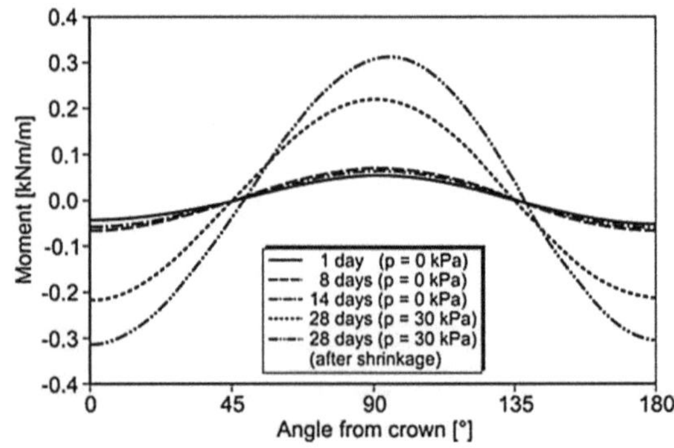

Fig. 7. Bending moment around rigid pipeline embedded in CLSM

5.2 PVC-U (Flexible) Pipe

Figures 8 and 9 show the distribution of contact stress between the pipe and the surrounding soil "CLSM" projected vertically to give the vertical stress component and horizontally to give the horizontal stress component acting along the flexible pipeline.

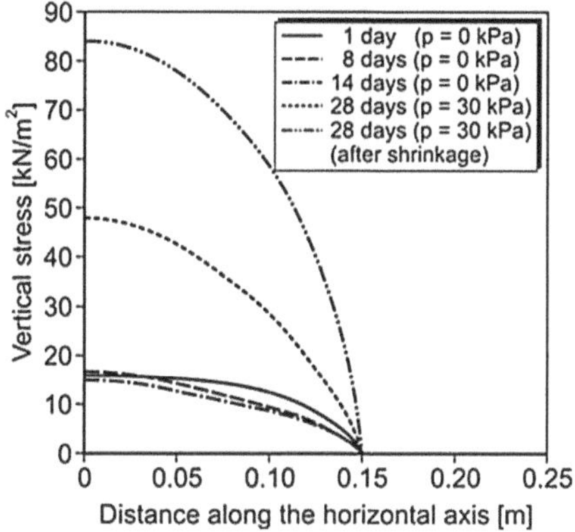

Fig. 8. Vertical stress around flexible pipeline embedded in CLSM

Similar as before, the vertical stress distribution (see Fig. 8) has its maximum value on the vertical (symmetry axis) of the pipe and the minimum value at the end of pipe (x = radius of the pipe = 0.15 m). In the first two weeks, the maximum value reached is 18.0 kN/m^2. Under the traffic loading, the maximum value reaches 48 kN/m^2 and by applying the shrinkage rate of 0.1% for CLSM elements, the vertical stress increases up to 82.0 kN/m^2.

Figure 9 shows the horizontal stress component drawn along the diameter of the pipeline (D = 0.30 m). The horizontal stress increases gradually to reach its maximum value of 18.0 kN/m^2 before shrinkage step was applied and then 48.0 kN/m^2 after the shrinkage was modeled. These values are smaller compared to the other values obtained by rigid pipe. However, both of them show that the hardening of self-compacting materials (CLSM) and the shrinkage of CLSM affect the contact stresses imposed on rigid as well as on flexible pipelines.

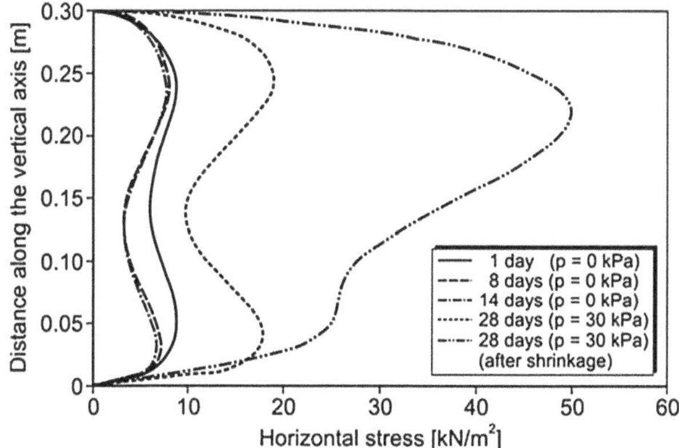

Fig. 9. Horizontal stress around flexible pipeline embedded in CLSM

The bending moment distribution is shown in Fig. 10. The bending moment up to 14 days is similar to the previous case (rigid pipe). After 28 days and the application of the vertical stress (30.0 kN/m^2) the bending moment increases up to 0.17 kNm/m, which is lower than the previous case (rigid pipe). The shrinkage induces an increase of the bending moment up to 0.23 kNm/m. In this case, the shrinkage causes an increase of the bending moment of almost 35%, which is smaller than in the case of rigid pipe.

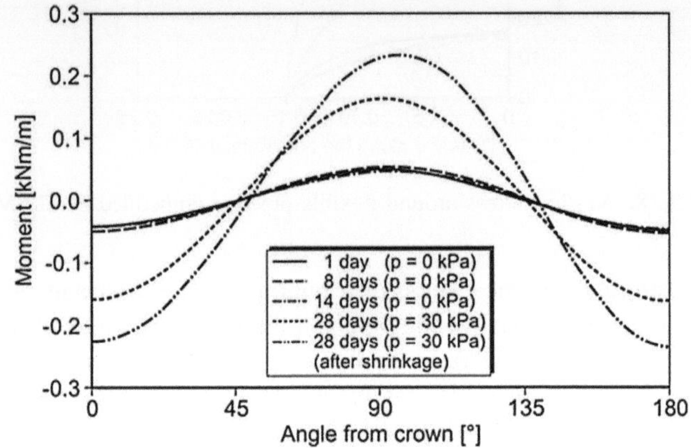

Fig. 10. Bending moment around flexible pipeline embedded in CLSM

6 Conclusions

For the investigations presented a FEM model was developed to simulate the behavior of pipelines embedded in self-compaction materials. The computations were executed taking the hardening process and shrinkage of self-compacting material into consideration. It is evident that the hardening process of the CLSM has as smaller effect on the bending moment compared to the effect of the shrinkage for both materials considered in the modelling process. However, both hardening and shrinkage of CLSM should be taken in account. In subsequent investigations, further parametric studies will be carried out for different trench widths and embedment depths.

References

Arsic, I.: Über die Bettung von Rohrleitungen in Flüssigböden. Institute for Soil Mechanics and Geotechnical Engineering, University of Bochum, Germany (2009). (in German)

Buhr, F.: Untersuchung zur Schwind- bzw. Schrumpfneigung von Flüssigböden im Zuge der Erhärtung. Bachelor thesis. IGtH, Hannover (2015). (in German). Unpublished

Bellaver, F.: Large diameter steel pipe field test using controlled low strength material and staged construction modeling using 3-D nonlinear finite element analysis. M.Sc. thesis, University of Texas at Arlington, Arlington, TX (2013)

Dezfooli, M., Abolmaali, A., Park, Y., Bellaver, F.: Staged construction modeling of steel pipes buried in CLSM using 3D nonlinear finite-element analysis. Int. J. Geomech. **15**(6), 1–13 (2015)

McGrath, T.J.: Pipe-soil interactions during backfill placement. Ph.D. thesis, University of Massachusetts, Amherst, MA (1998)

Simulia: User's Manual, Version 6.13. Simulia, Providence (2015)

Suleiman, M.T., Lohnes, R.A., Wipf, T.J., Klaiber, F.W.: Analysis of deeply buried flexible pipes. Transp. Res. Rec. **1849**, 124–133 (2002)

Application of Vacuum Consolidation for the Improvement of Tunis Soft Soil

Halima Jebali[✉], Wissem Frikha, and Mounir Bouassida

Ecole Nationale d'Ingénieurs de Tunis, LR14ES03-Ingénierie Géotechnique, Université de Tunis El Manar, BP 37 Le Belvédère, 1002 Tunis, Tunisia
hlm.jebali@gmail.com,
{wissem.frikha, mounir.bouassida}@enit.utm.tn

Abstract. An experimental investigation was carried out on the reconstituted Tunis soft soil that was extracted from the center of Tunis City at 35 m depth. A specific laboratory test apparatus was designed and manufactured to perform three consolidation series of tests by applying a negative pressure, a preload or a vacuum pressure combined with preload. Excess pore pressure and settlement were measured during the consolidation process. Experimental results showed that for the same magnitude of preload and the vacuum pressure of 4 kPa, 8 kPa and 16 kPa and 30 kPa, the settlement caused by the vacuum pressure is lower than that generated by the preload. While the settlement generated by preloads of 60 kPa and 100 kPa are slightly smaller than those caused by the vacuum consolidation of 60 kPa and 100 kPa. Recorded excess pore pressure around the geodrain dissipated more quickly compared to excess pore pressure close to cell border. The normalized excess pore pressure for preload decreases when the applied preload increases and then increases quasi-linearly with the increase of horizontal distance between transducers to geodrain. While, the normalized excess pore pressure for vacuum pressure increases with the increase of the applied negative pressure and decreases with the increase of horizontal distance between transducers and the geodrain.

Keywords: Vacuum pressure · Preload · Settlement · Excess pore pressure

1 Introduction

Preloading, usually implemented by an embankment, is the simplest technique to improve the undrained shear strength of soft soils. However, the main disadvantage of preloading, regardless its cost, remains the long time to attain the expected improvement. Hence the recourse to vertical drains revealed necessary for infrastructure projects in Tunisia to accelerate the consolidation of soft soils (Bouassida and Hazzar 2008). For thirty years, the technique of vacuum consolidation associated with vertical drains found successes all over the world (Mohamed Elhassan and Shang 2002; Chai et al. 2005; Indraratna et al. 2009; Saowapakpiboon et al. 2009; Indraratna et al. 2012).

In this paper, an experimental study was carried out on the reconstituted Tunis soft soil (TSS) improved by geodrain (Mebradrain 88). At the knowledge of authors, it is the first contribution in studying the improvement of TSS by geodrain combined with

© Springer International Publishing AG 2018
M. Bouassida and M.A. Meguid (eds.), *Ground Improvement and Earth Structures*, Sustainable Civil Infrastructures, DOI 10.1007/978-3-319-63889-8_8

vacuum consolidation technique either for research purpose or for real project application in Tunisia. A specific laboratory test apparatus was designed and manufactured to run the consolidation tests by applying vacuum pressure, or preload or vacuum pressure combined with preload.

A detailed characterization of the soil, the equipment used and the experimental procedure are presented. The experimental results of the evolution of excess pore pressure and settlement versus time of all series test are discussed and interpreted. This experimentation was conducted at the geotechnical engineering research laboratory of National Engineering School of Tunis.

2 Experimental Procedure

2.1 Preparation of Samples

The soil used for the present experimental investigation is a remolded TSS obtained from the center of Tunis City at a depth of 35 m. The extracted sample has greyish green color with a characteristic smell and contains shells. A grain size analysis performed by hydrometer and sieve methods in accordance with standards, NF P 94-056 (AFNOR 1995a) and NF P 94-057 (AFNOR 1995b) was carried out on extracted specimens. Tunis soft soil (TSS) contains 98% of particles with dimensions lower than 80 μm and includes 32% of clay and 30% of silt (Jebali et al. 2013). Atterberg limits' tests were performed according to the NF P 94-051 (AFNOR 1995c): liquid limit w_l is equal to 65% and plastic limit w_p is equal to 37.7%.

The extracted specimens were subjected to a reconstitution procedure by eliminating shell debris to obtain a homogeneous soil.

The reconstitution of TSS consisted of wet sieving the natural soil through the 100 μm sieve. The sieved soft soil is air-dried until water content reached about 1.4 times its liquid limit (w = 91%).

2.2 Used Equipment

A specific laboratory test apparatus was designed and manufactured to apply the vacuum pressure, a preload or a vacuum pressure combined with a preload. For the design of the test apparatus, the main required criteria were the sealing of the system, the measurement of the excess pore pressure and the settlement along with the performing test.

The test apparatus consisted of two consolidometers, an acquisition system composed of 4 pore pressure Transducers PPT_1; PPT_2; PPT_3 and PPT_4, two displacement transducers, a data acquisition card, a multiplexer and a computer provided with a data acquisition program. The first consolidometer was used for consolidation tests by preload, whereas the second one was used for the vacuum consolidation tests (Fig. 1).

2.3 Testing Procedure

The first saturated porous stone covered by wet filter paper was placed at the lower base of soil specimen. The pore pressure transducers were placed in straws of 5 mm diameter

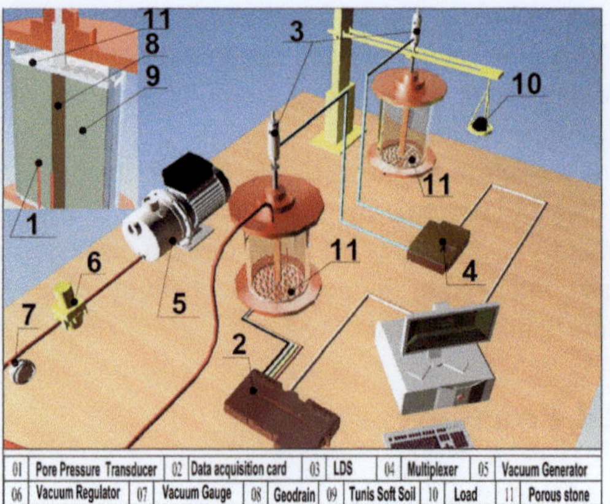

| 01 | Pore Pressure Transducer | 02 | Data acquisition card | 03 | LDS | 04 | Multiplexer | 05 | Vacuum Generator |
| 06 | Vacuum Regulator | 07 | Vacuum Gauge | 08 | Geodrain | 09 | Tunis Soft Soil | 10 | Load | 11 | Porous stone |

Fig. 1. Schematic of testing apparatus

and 100 mm height. The straws were installed in the holes of porous stone in preselected locations (Fig. 2). The PPT_1, PPT_2, PPT_3 and PPT_4 were placed at a distance of 1.5 cm, 3 cm, 4.5 cm and 5.5 cm, respectively from the center of the porous stone. Thereafter, the inner cell was lubricated with grease and the reconstituted TSS was placed into consolidometer by sub-layers. Those sub-layers, each of mass about of 500 g, were

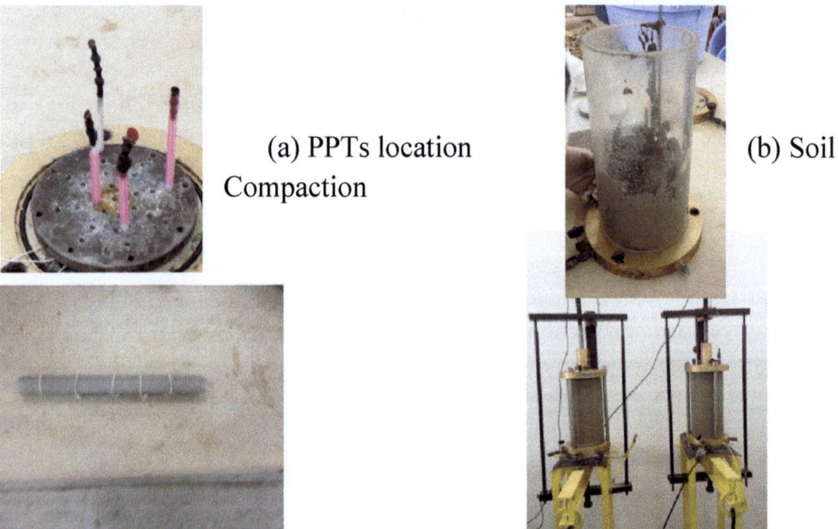

(a) PPTs location Compaction

(b) Soil

(c) Mebradrain 88 geodrain

(d) consolidation cells

Fig. 2. Sequences of preparation and performing consolidation test

gently compacted with a thin rod by applying 25 blows to remove air bubbles. The soil was gently placed using a spoon to avoid any damage of transducers. As soon as the soil has reached the height of the cell of 29 cm, a portion of geodrain (width of 18 mm and thickness of 5 mm) was inserted into the middle of the reconstituted specimen. Then, the second saturated porous stone covered with wet filter paper was placed at the top of specimen; a mass of 15 kg was placed until a settlement of 4 cm was recorded. Then the upper side of cell was closed and the test is ready to start.

3   Conducted Tests

Three series of tests were carried out to study the consolidation of TSS by applying a preload (Series 1), a vacuum pressure (Series 2) and a vacuum pressure combined with a preload (Series 3). Table 1 summarizes the series of tests performed on TSS improved by vacuum and preload techniques.

Table 1. Summary of conducted tests

	Tests name	Preload, q (kPa)	Vacuum pressure, p_v (kPa)	Number of tests
Series 1	SL-4-100-G	4, 8, 16, 30, 60, 100	0	2
Series 2	VC-4-100-G	0	4, 8, 16, 30, 60, 100	2
	VC-100-G	0	100	1
Series 3	SL-50-VC-50-G	50	50	1

The first series included a consolidation test by preload (SL). In SL-4-100-G test, the soft soil specimen improved by geodrain (G) was subjected to preloads of 4 kPa, 8 kPa, 16 kPa, 30 kPa, 60 kPa and 100 kPa. In this case, each preload level was kept constant during 10 days at minimum to attain a stabilized settlement.

The second series comprised two vacuum consolidation tests (VC). In VC-4-100-G test, the reconstituted soft soil specimen was subjected to negative pressure of 4 kPa, 8 kPa, 16 kPa, 30 kPa, 60 kPa and 100 kPa. Each vacuum pressure was maintained constant for at least 10 days. In regard to VC-100-G test, a negative pressure of 100 kPa was applied to achieve a constant settlement. In the third series, the SL-50-VC-50-G test was performed on the reconstituted TSS specimen subjected to a negative pressure of 50 kPa combined with preload of 50 kPa. The repetition of tests was carried out with the same procedures to check their repeatability and to validate the experimental procedure and used equipment along with the apparatus measure.

4   Results and Interpretations

This section will focus on the variations of settlements and excess pore pressure (Δu) generated by applied preload (q) or vacuum pressure (p_v) or vacuum pressure combined with preload. The obtained results from performed tests are presented and discussed.

4.1 Settlement

Figure 3 illustrates the evolution of recorded settlement versus time for the SL-4-100-G test (Series 1), VC-4-100-G, VC-100-G tests (Series 2) and SL-50-VC-50-G test (Series 3). A maximum settlement of 59.46 mm recorded for VC-100-G test (Series 2) when a negative pressure of 100 kPa is subjected to soft soil specimen. The final primary consolidation settlement during SL-50-VC-50-G test (Series 3) subjected to negative pressure of 50 kPa combined with a preload of 50 kPa reached 60.46 mm. The final primary consolidation settlement for the VC 4-100-G test (Series 2) induced by successive vacuum pressures (from 4 kPa to 100 kPa) reached 57.41 mm. In turn, SL-4-100-G test (Series 1) subjected to preloads (from 4 kPa to 100 kPa) generated a final primary consolidation settlement of 58.13 mm. Considering all these results, it can be noted that vacuum pressure of 50 kPa combined with preload of 50 kPa generated greater settlement than those recorded during VC-100-G test. The settlement obtained from SL-50-VC-50-G test reached its peak value after 7.5 days while the final settlement obtained during the VC-100-G test is reached after 9 days. The settlement obtained from VC-4-100-G and SL-4-100-G tests, under incremental preloads and negative pressures of 4 kPa up to 100 kPa, are attained after 60 days. Therefore, it follows that the major benefit of combining the vacuum pressure with preloading is the gain in time of the settlement stabilization that reflects the end of the consolidation process.

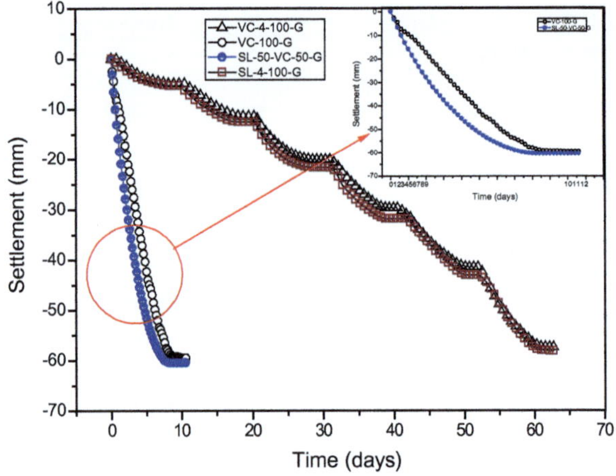

Fig. 3. Settlement versus time for VC-100-G, VC-4-100-G, SL-4-100-G and SL-50-VC-50-G tests.

4.2 Excess Pore Pressure

Two ratios δ_1 and δ_2 are introduced to study the effects of preload and vacuum pressure on the variation of excess pore pressure.

δ_1 is the normalized excess pore pressure for preload (q):

$$\delta_1 = \frac{\Delta u_{\max}}{q} \tag{1}$$

δ_2 is the normalized excess pore pressure for vacuum pressure (p_v):

$$\delta_2 = \frac{\Delta u_{\max}}{p_v} \tag{2}$$

Figure 4 well illustrates, for a given preload that the excess pore pressure increases as the distance from central geodrain to the pore pressure transducers also increases. This explains the accelerated consolidation of reconstituted soft clay the more it is closed to the geodrain. From figure it also seen that the applied preload becomes higher the trend of generated excess pore pressure becomes almost uniform. Therefore the increase in induced excess pore pressure as function of the horizontal distance from geodrain to the cell border becomes less pronounced when the applied preload decreases.

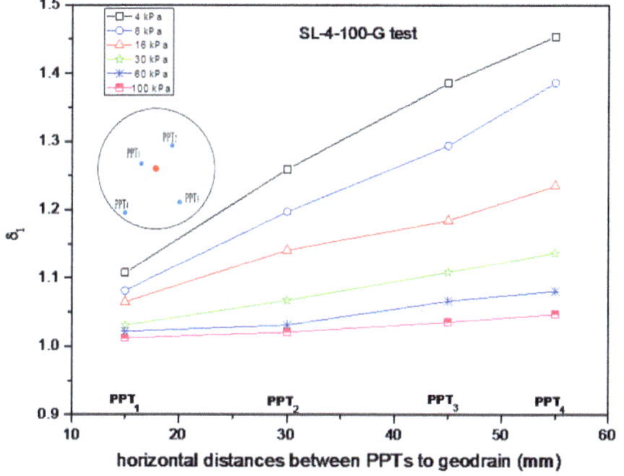

Fig. 4. Variation of δ_1 (the normalized excess pore pressure for preload) versus horizontal distances of pore pressure transducers from geodrain for SL-4-100-G test.

The variation of δ_2 versus the horizontal distances between pore pressures transducers to geodrain obtained from the VC-4-100 G-test (Series 2) is illustrated in Fig. 5. It should be noted that the ratio δ_2 increases with the increase of the applied negative pressure and decreases with the increase of horizontal distance between transducers to geodrain. Therefore the decrease of generated negative excess pore pressure as function of the horizontal distance from geodrain to the cell border is noted when the applied vacuum pressure increases.

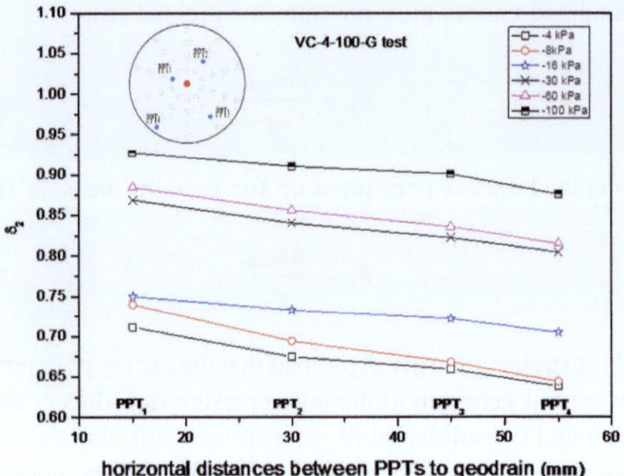

Fig. 5. Variation of δ_2 (the normalized excess pore pressure for vacuum pressure) versus horizontal distances of pore pressure transducers from geodrain for VC-4-100-G test.

5   Conclusions

A first evaluation of improvement technique using the vacuum consolidation to accelerate the consolidation of Tunis soft soil was presented herein.

An experimental laboratory investigation was carried out on reconstituted TSS improved by geodrain in three configurations.

Experimental results have shown that for the same magnitude of preload and the vacuum pressure less than or equal to 30 kPa, the settlement caused by the vacuum pressure is less than that generated by the preload. While the settlement generated by preloads of 60 kPa and 100 kPa are slightly smaller compared to those caused by vacuum consolidation of 60 kPa and 100 kPa. These results have shown that the end of consolidation process is explained by the dissipation of excess pressure and stabilization of soil deformation. Recorded excess pore pressure around the geodrain dissipates more quickly compared to excess pore pressure close to cell border.

The combination of vacuum pressure with preloading revealed an effective method for accelerating soft soil consolidation during which the application of negative pressure soil generating an inward horizontal deformation.

References

AFNOR, NF P 94-056: Sols: reconnaissance et essais. Analyse granulométrique par tamisage et Analyse granulométrique par sédimentométrie. French standard (1995a)

AFNOR, NF P 94-057: Sols: reconnaissance et essais. Analyse granulométrique par sédimentométrie. French standard (1995b)

AFNOR, NF P 94-051: Sols: reconnaissance et essais. Détermination des limites d'Atterberg. Limite de liquidité à la coupelle- Limite de plasticité au rouleau. French standard (1995c)

Bouassida, M., Hazzar, L.: Comparison between stone columns and vertical geodrains with preloading embankment techniques. In: Prakash, S. (ed.) Proceedings of the 6th International Conference on Case Histories in Geotechnical Engineering and Symposium in Honor of Professor James K. Mitchell, Arlington, Virginia, USA, 11–16 August 2008

Chai, J.C., Carter, J.P., Hayashi, S.: Ground deformation induced by vacuum consolidation. Geotech. Geoenviron. Eng. **131**(12), 1152–1561 (2005). ASCE

Jebali, H., Frikha, W., Bouassida, M.: Assessment of Carillo's theory for improved Tunis soft soil by geodrains. In: Proceedings of the ICSMGE 2018, Paris, 2–6 September 2013, pp. 2509–2512 (2013)

Indraratna, B., Rujikiatkamjorn, C., Ghandeharioon, A.: Modeling of soft ground consolidation via combined surcharge and vacuum preloading. In: Karstunen, M., Leoni, M. (eds.) Geotechnics of Soft Soils – Focus on Ground Improvement, pp. 43–53 (2009)

Indraratna, B., Rujikiatkamjorn, C., Balasubramaniam, A.S., Mcintosh, G.: Soft ground improvement via vertical drains and vacuum assisted preloading. Geotext. Geomembr. **30**, 16–23 (2012)

Mohamed Elhassan, E., Shang, J.Q.: Vacuum and surcharge combined one-dimensional consolidation of clay soils. Can. Geotech. J. **39**(5), 1126–1138 (2002)

Saowapakpiboon, J., Bergado, D.T., Youwai, S., Chai, J.C., Wanthong, P., Voottipruex, P.: Measured and predicted performance of prefabricated vertical drains (PVDs) with and without vacuum preloading. Geotext. Geomembr. **28**(1), 1–11 (2009)

Numerical and Experimental Studies
of Sand-Clay Interface

Wissem Frikha[✉] and Belgacem Jellali

Ecole Nationale d'Ingénieurs de Tunis, LR14ES03-Ingénierie Géotechnique,
Université de Tunis El Manar, BP 37 Le Belvédère, 1002 Tunis, Tunisia
Frikha_wissem@yahoo.fr, belgacem.jellali@gmail.com

Abstract. A recent experimental work of Frikha et al. (2015) demonstrated that when the number of granular columns increases at a fixed area replacement ratio "η", the contact surface between the reinforcing columns and the clayey soil increases and results in an increase in the soil-column interface friction, which enhances the strength properties of the reinforced soft clayey soil. Granular Columns/soil interface strength can play then an important role in the bearing capacity of reinforced soil. However, it would appear that none of the methods proposed of design of granular column to date are able to reliably predict performance to a satisfactory level of accuracy due to the intrinsic characteristics of interface properties Columns/soil.

The present paper try to carry out into the shear behavior of sand/clayey soil interface using a simple direct shear test. In the shear test a shearing surface is produced and the resistance to shearing is defined as the shear resistance. The frictional resistance is, however, determined on an existing joint surface. The frictional behavior at the soil – other materials interface is usually and commonly obtained from direct shear tests. A numerical modelling analysis of direct shearing box test using FLAC3D was performed in order to define the interface parameter by comparing the experimental results with the numerical one. The principal purpose of this work is to define the properties of the interface for a future numerical simulation of behavior of clayey soil reinforced by granular column.

1   Introduction

Modeling of soil-structure interaction is very important in geotechnical engineering including hydraulic structures. It is relevant to a wide range of project problems, such as numerical analysis of concrete-faced rock fill dams, retaining walls, shallow foundations, piles, tunnels, reinforced earthworks, and geosynthetic liners. The relationship between shear stress and shear displacement of the soil-structure interface plays a major role in modeling soil-structure interactions (Wu et al. 2011).

Several investigations were conducted for determining the bearing capacity of an isolated column by considering several methods (Bouassida and Hadhri 1995) including the state of stress (Aboshi et al. 1979), a failure mechanism combined with a state of stress (Greenwood 1970; Hughes et al. 1975; Datye 1982; Balaam and Booker 1985; Frikha and Bouassida 2015), or a failure mechanism only is used (Bouassida and Jellali 2002).

© Springer International Publishing AG 2018
M. Bouassida and M.A. Meguid (eds.), *Ground Improvement and Earth Structures*,
Sustainable Civil Infrastructures, DOI 10.1007/978-3-319-63889-8_9

Hu (1995) built a laboratory-scale model to examine the behavior of a cohesive soft soil reinforced by a group of granular columns supporting a rigid footing. He found that reinforcement by a group of columns is more effective than that by an isolated column. The interaction between the columns and the soft clay was found to efficiently contribute to the enhancement of the load bearing capacity and to provide a wider transfer of loading. Bachus and Barksdale (1984) concluded that there is only a slight increase in the ultimate load bearing capacity per column when the number of columns increases. However, in their tests, the effect of lateral confinement was more significant since the reinforcing columns were set close to the borders of the testing box. Wood et al. (2000) commented on the tests performed by Hu (1995) and confirmed that the mode of failure for each column, in the case of group-column reinforcement, depends on its location within the group, its length, and the type of loading. Their results showed that the pre-failure mechanisms and the failure modes of a granular-column group are different from those observed for a single column. They reported that the area replacement ratio affects the extent of the columns' interaction and the load transferred to the soft clay in between the columns. This research concluded that a significant improvement in the bearing capacity depends on a minimum area replacement ratio of 25%.

A recent experimental work performed by Frikha et al. (2015) demonstrate that when the number of columns increases at a fixed area replacement ratio "η", the contact surface between the reinforcing columns and the clayey soil increases and results in an increase in the soil-column interface friction, which enhances the strength properties of the reinforced soft clayey soil.

Granular columns/soil interface strength can play then an important role in the bearing capacity of reinforced soil. However, it would appear that none of the methods proposed of design of granular column to date are able to reliably predict performance to a satisfactory level of accuracy due to the intrinsic characteristics of interface properties granular columns/soil.

The present paper tries to carry out into the shear behavior of granular Column/Clayey soil interface using a simple direct shear test (Fig. 1). In the shear test a shearing surface is produced and the resistance to shearing is defined as the shear resistance. The frictional resistance is, however, determined on an existing joint surface. The frictional behavior at the soil– other materials (e.g. geosynthetic, geogrid, tire shreds, rubber chips, geofoam, polymer, etc.) interface is usually and commonly obtained from direct shear tests (O'Rourke et al. 1990; Bernal et al. 1997; Lee and Manjunath 2000; Xenaki and Athanasopoulos 2001; Bergado et al. 2006; Liu et al. 2009; Palmeira 1988; Anubhav and Basudhar 2010).

The principal purpose of this work is to define the properties of the interface for a future numerical simulation of behavior of clayey soil reinforced by granular column. There are in fact several instances in geomechanics in which it is desirable to represent planes on which sliding or separation can occur. Include for example, the cases of joint, fault or bedding planes in a geologic medium; an interface between a foundation and

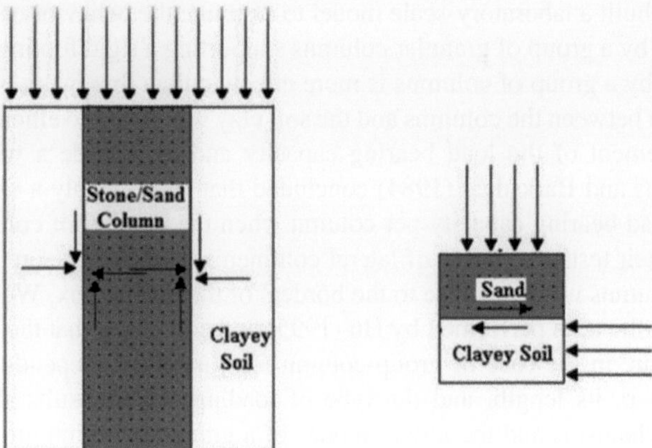

Fig. 1. Proposal model

the soil; a contact plane between a bin or chute and the material that it contains; a contact between two colliding objects; and a planar "barrier" in space, which represents a fixed, non-deformable boundary at an arbitrary position and orientation.

The sample preparation, test method, and test results of an experimental program of direct shear testing in the interface between two layers of sand/clayey are undertaken. A simulation of performed model using finite-difference code (Itasca FLAC3D) is also conducted and interpreted.

2   Studied Soil

The tested soft soil is grey in color, contains shell debris, and has a characteristic smell. The properties of Tunis soft soil are a liquid limit of $\omega_l = 66.5\%$, a plastic limit of $\omega_p = 23.9\%$, a plasticity index of IP = 42.6%, a natural water content of $\omega = 74\%$, a compression index of $C_c = 0.64$, and an oedometer modulus of E = 1700 kPa. According to the French Standard XP P 94-011 classification, the tested soft soil is classified as a highly plastic clay.

The material used is siliceous sand classified as CEN in accordance with the European standards EN 196-1. The sand particles are generally isometric and rounded in shape. The coefficient of uniformity of the CEN sand is $C_u = 6$ and its dry unit weight is $\gamma_d = 16$ kN/m^3.

The grain size distributions of Tunis soft soil and CEN Standard sand are depicted in Fig. 1. It is noticed that 60% of the particles of Tunis soft soil are smaller than 2 μm and 98.3% are smaller than 80 μm. As seen in Fig. 2, the mean particle diameter, D_{50}, of the CEN Standard sand is 0.7 mm.

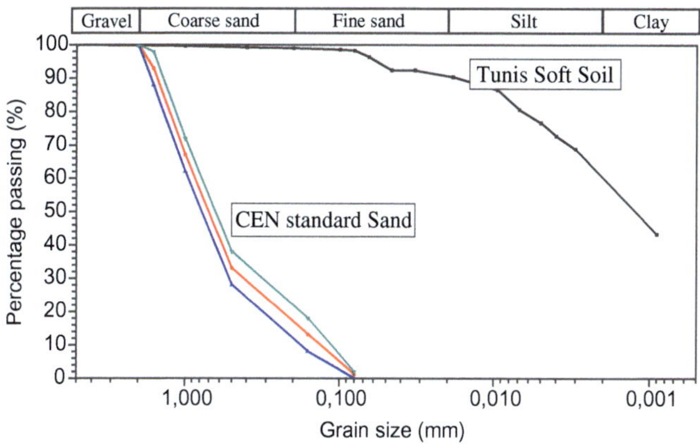

Fig. 2. Grain size distribution of used clayey and sandy soil

3   Experimental Procedure

The first step consists of wet sieving the natural soft soil through a 100-µm sieve. Then, the sieved soft soil is air-dried until its moisture content reaches 100%, which represents approximately 1.5 ωl.

The second step is to subject the slurry material to a vertical pre-consolidation pressure of K_0 in a special mould (consolidometer container). The consolidometer is a metallic box, 6 × 6 cm in surface and 20 mm in height (Fig. 3).

The inner surface of the consolidometer is lubricated with silicone grease prior to the slurry placement in order to reduce the effects of roughness, and therefore, allow an easy unmolding of the specimens. The inner surface of the consolidometer is also lined with filter paper to accelerate the consolidation. The slurry is then carefully placed into the mould, using a spoon, where it is frequently tamped with a plastic tamper to avoid the formation of air during the filling phase. When the predetermined specimen height

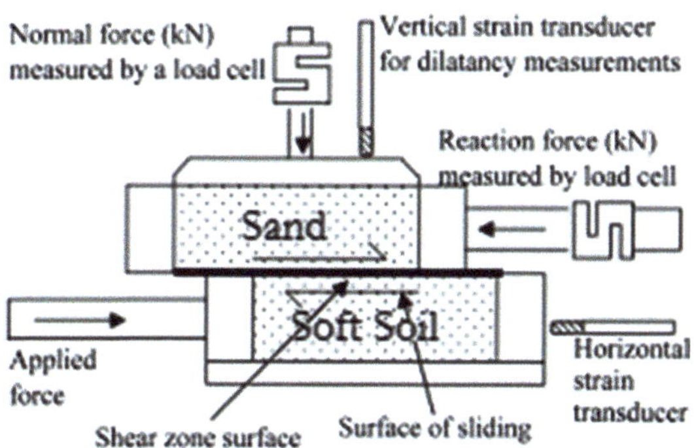

Fig. 3. Used shear box

is attained, the consolidation loading cap is placed to start the loading phase. The specimens are consolidated under a vertical stress of 54 kPa for at least seven days. At the end of the consolidation phase, the specimen height is 1,2 cm. During the pre-consolidation phase, drainage is allowed at the top and bottom of the specimen through the installed upper and lower porous stones. A dial gauge on the loading cap is used to measure the axial deformation of the tested specimen. The consolidation is considered to be complete when quasi-constant axial deformation is recorded. The pressure is then reduced to zero and the loading cap is removed. The soft clay sample is placed in the shear lower half-box.

The sample is then cut at the shear plane of lower half-box by means of a steel wire to obtain a final height of 1 cm. the upper half-box is then fixed so as to secure the two half boxes and is filled with a dry sand in small layers (20 gr of sand per layer), in order to attain the final height of sand layer (10 mm). Each layer is compacted by a plastic tamper. The same weight (20 gr) of dense compacted sand is considered in all tests.

The obtained sample composed of two layers is then subjected to direct shear stress in a shear box equipment (Fig. 3) under a velocity of $V = 0.5$ mm/min and a normal stresses of $\sigma_N = 54.8$ kPa, 67.4 kPa and 80.0 kPa.

4   Experimental Results

Figures 4 present typical results of the direct shear box tests carried out on (i) reconsolidated soft soil, (ii) normalized sand and (iii) sample composed from two layers (Sand and Clay) subjected to different normal stresses. These figures show the variations in shear stress, $\tau = F/S$, with respect to the calculated linear strain under undrained conditions. Linear strain ε_a is calculated as the measured displacement divided by the specimen's initial length $\varepsilon_a = \delta l/l$ (%).

For the case of the Tunis soft soil, shear stresses are quasi-independent of normal stresses.

Figure 5 show the obtained results of shear stress vas a function of applied Normal stress for (i) Clay-Clay (ii) Sand-Sand and (ii) Clay-Sand. In the three cases the shear stress is evolution quasi-linear with the normal stress. The obtained cohesion "c" is 34.0, 23.7 and 0 for Clay-Clay, Sand-Sand and Clay-Sand respectively. Whereas the friction angle "φ" is 9.44°, 27.6° and 43.1° for Clay-Clay, Sand-Sand and Clay-Sand respectively. Both the cohesion and the friction angle derived from performed test on interface clay-sand are between those obtained for clay-clay and sand-sand interfaces.

5   Numerical Modelling and Analysis

FLAC3D provides interfaces that are characterized by Coulomb sliding and/or tensile and shear bonding. Interfaces have the properties of friction, cohesion, dilation, normal and shear stiffness, and tensile and shear bond strength. FLAC3D represents interfaces as collections of triangular elements (interface elements), each of which is defined by three nodes (interface nodes) (Figs. 6, 7 and 8).

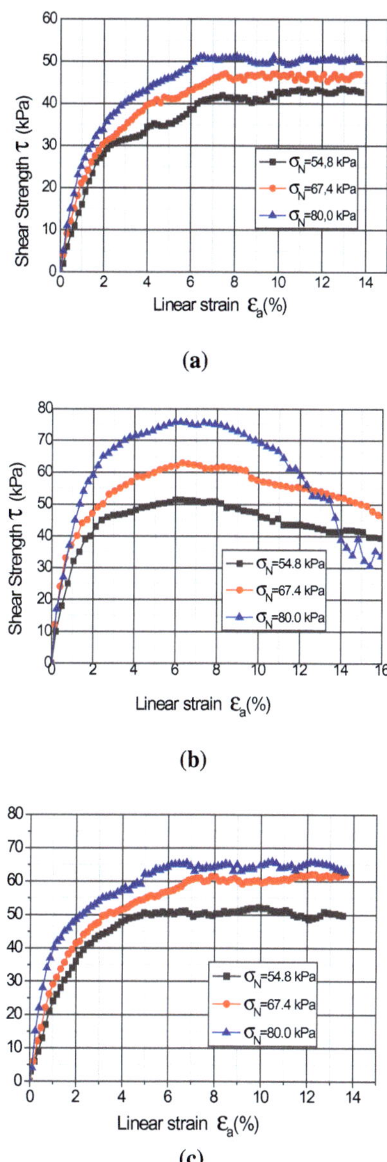

Fig. 4. Results of shear tests performed on (a) Clay-Clay (b) Sand-Sand and (c) Clay-Sand

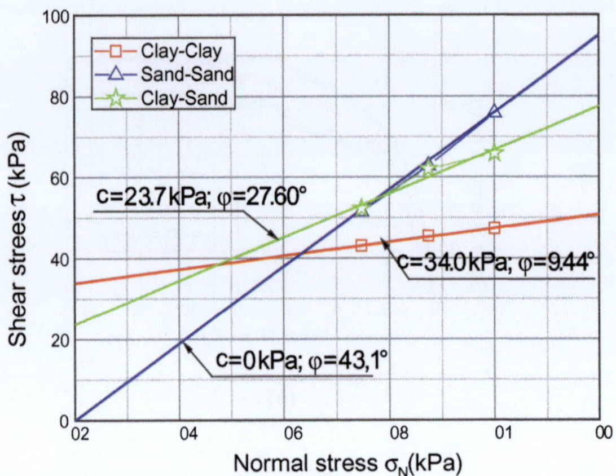

Fig. 5. Shear stress versus normal stress for (i) Clay-Clay (ii) Sand-Sand and (ii) Clay-Sand

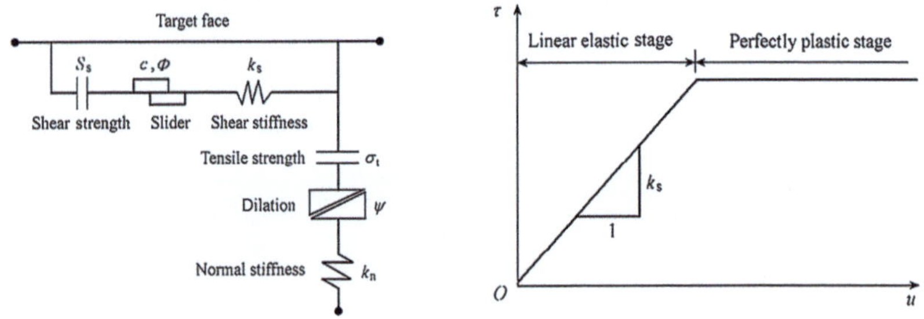

Fig. 6. Theory of interface element implemented in FLAC3D

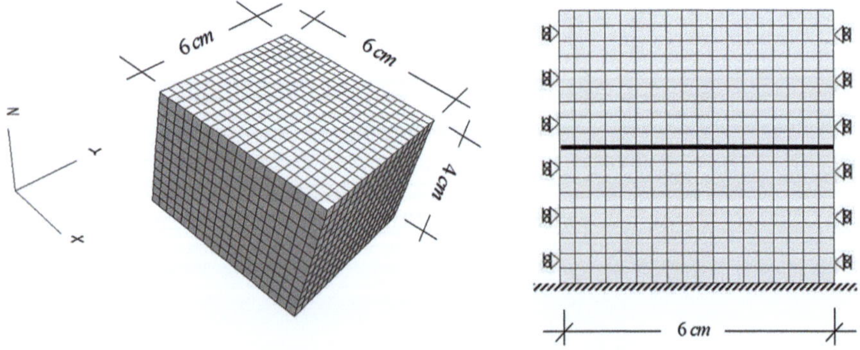

Fig. 7. Grid of used numerical model

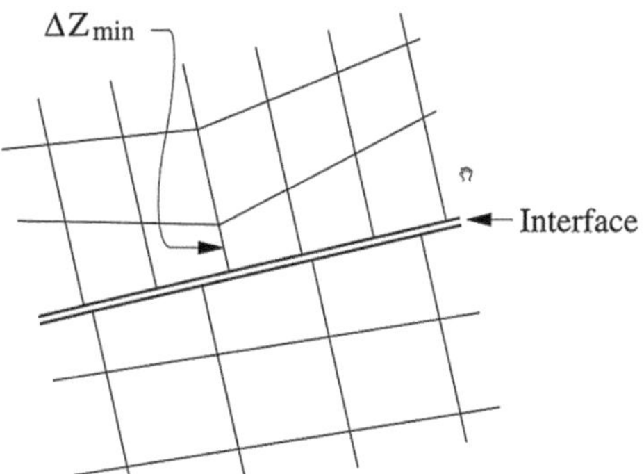

Fig. 8. Zone dimensions used in stiffness calculation

The normal and shear forces that describe the elastic interface stage response are determined at calculation time $(t + \Delta t)$ the following equations provided by FLAC3D (FLAC3D Online Manual 2005):

$$F_n^{(t+\Delta t)} = (k_n u_n + \sigma_{n0})A, \; F_{si}^{(t+\Delta t)} = \left(F_{si}^{(t)} + k_s.\Delta u_{si}^{(t+\Delta t/2)} + \sigma_{si}\right)A \tag{1}$$

Where $F_n^{(t+\Delta t)}$ is the normal force at the calculation time $t + \Delta t$; k_n is the normal stiffness of interface element; u_n is the normal penetration of the interface node into the target face; σ_{n0} is the normal effective stress added due to interface stress initialization; A is the representative area associated with the interface node; $F_{si}^{(t+\Delta t)}$ and $F_{si}^{(t)}$ are the shear forces at calculation times $t + \Delta t$ and t respectively; k_s is the shear stiffness of interface element, which is a constant in the elastic stage; $\Delta u_{si}^{(t+\Delta t/2)}$ is the incremental shear displacement between $t + \Delta t$ and t; and σ_{si} is the vector of additional shear stress due to interface stress initialization (FLAC3D Online Manual 2005).

According to the Mohr-Coulomb criteria, the yield relationships in the shear and normal directions are:

$$F_{Smax} = CA + Tan\varphi(F_n - pA), \; F_n = \sigma_t \tag{2}$$

where F_s max is the maximum shear strength, c is the cohesion of the interface, φ is the friction angle of the interface, F_n is the normal force, p is the pore pressure, and σ_t is the normal tensile strength of the interface.

If the interface does not have dilation characteristics, the forces are corrected as follows:

$$F_{si} = F_{s\,max} \tag{3}$$

If the interface has dilation characteristics, the forces are corrected as follows (FLAC3D Online Manual 2005):

$$F_n = F_n + \frac{(|F_s|_0 - F_{smax})}{Ak_s} tan\psi \, k_n, \, F_{si} = F_{s\,max} \tag{4}$$

Where $|F_s|_0$ is the shear force before the above corrections are made, and ψ is the dilation angle of the interface.

A series of numerical analysis has been conducted to simulate the direct shear tests using FLAC3D code by ITASCA. The model geometry is shown in Fig. 4. The numerical model of the test is composed of two parts: each part is a shear box with soil in it and discretized 5832 brick-shaped mesh elements. Soil strengths are defined by the Mohr-Coulomb failure criterion.

The interface between the dissimilar materials is modeled as linear spring-slider system with interface shear strength defined by a linear-elastic perfectly plastic model with a Mohr-Coulomb failure criterion. The relative interface movement is controlled by interface normal stiffness (k_n) and shear stiffness (k_s). A recommended thumb rule is that k_s and k_n be set to ten times the equivalent stiffness of the stiffest neighboring zone. The maximum stiffness value is given by as:

$$k_n = k_s = 10 \times max \left[\frac{K + \frac{4}{3}G}{(\Delta z)_{min}} \right] \tag{5}$$

Where the parameters $(\Delta z)_{min}$, K and G are the smallest dimensions in normal direction, bulk modulus and shear modulus continuum zone adjacent to the interface respectively. This approach gives the preliminary values of the interface stiffness components, and these can be adjusted to avoid intrusion to adjacent zone and to prevent excessive computation time. The interface behavior of clay-Sand is represented numerically at each node by a rigid attachment in normal direction and spring-slider in the tangent plane to the interface surface. The required input parameters for interaction between interface grids are: (i) coupling spring cohesion (ii) coupling spring friction (iii) coupling spring stiffness. Table 1 summarizes the geotechnical parameters of each studied soil.

Table 1. Material properties used in numerical simulation

Material	Cohesion c (kPa)	Friction φ (°)	Bulk modulus K (MPa)	Shear modulus G (MPa)
Sand	0	43.38	0.29167	0.14615
Clay	34.35	10	0.23167	0.024138

Table 2. Interface properties used in numerical simulation

Interface	$k_n = k_s$ (N/m)	Friction φ (°)	Cohesion c (kPa)	Tensile (kPa)
Sand-Sand	424.04 10^7	43.38	0	10^{10}
Sand-Clay	424.04 10^7	27.6	23.75	10^{10}
Clay-Clay	237.47 10^7	10	34.35	10^{10}

In accordance with Eq. 5, parameters of different interfaces are listed in Table 2.

The analysis was carried out considering first step, only normal stress was applied on the top surface of the model and in the second step, shear stress was applied. All of the analyses were performed using normal stress of 54.8, 67.4 and 80.0 kPa respectively.

Figure 9 shows the boundary condition model. The whole model is restrained in the y direction while the bottom part is also restrained in z directions at its base.

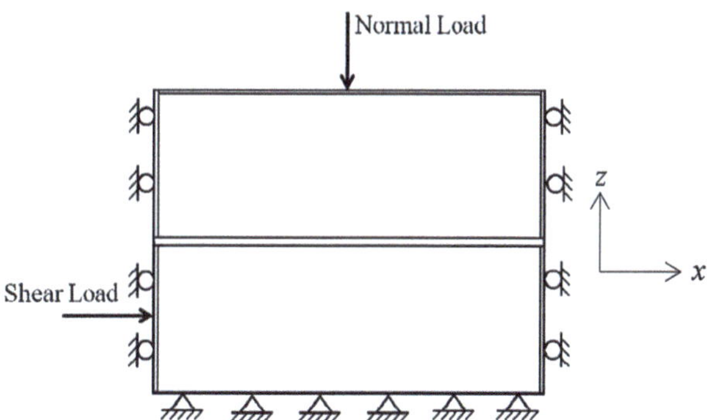

Fig. 9. Boundary conditions in xz plane

In the initial step, a normal load is applied at the top of the model making a set of stresses representing soil initial state installed in the grid, and then FLAC3D is run under elastic assumptions until an equilibrium state is obtained. The initial horizontal stress is related to the initial vertical stress by at rest lateral earth pressure coefficient (K_0) which equals to $(1 - \sin \varphi)$, where φ is the angle of internal friction of the soil. In the second step a horizontal velocity of 5×10^{-7} m/s is applied to the bottom box in x direction.

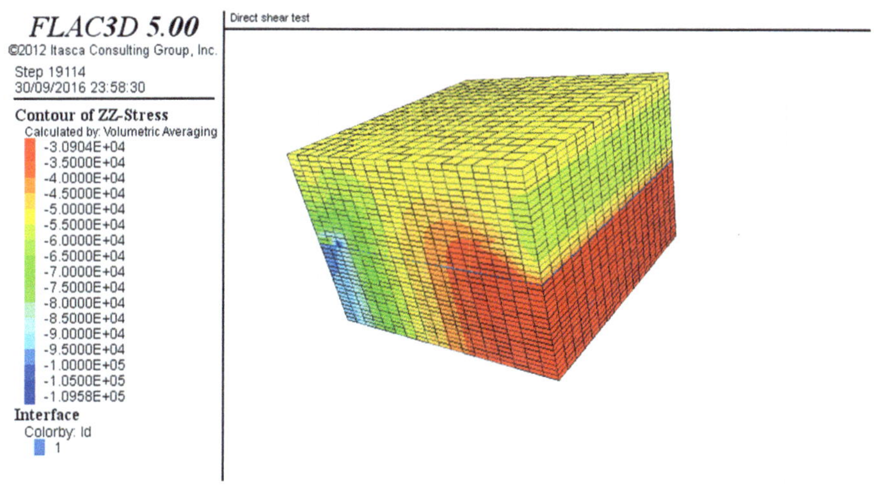

Fig. 10. Vertical stress obtained at the end of simulation

In this present simple analysis of the last section, the friction in the interface is assumed to be fully mobilized. In fact, the degree of mobilization will depend on the shear displacement.

Figure 10 presents an example of obtained vertical stress at the end of simulation numerical modelling. Numerical results are shown in Figs. 11 together with the experimental results. It can be seen that the model is able to reproduce the behavior of soil-soil interface with different normal stresses.

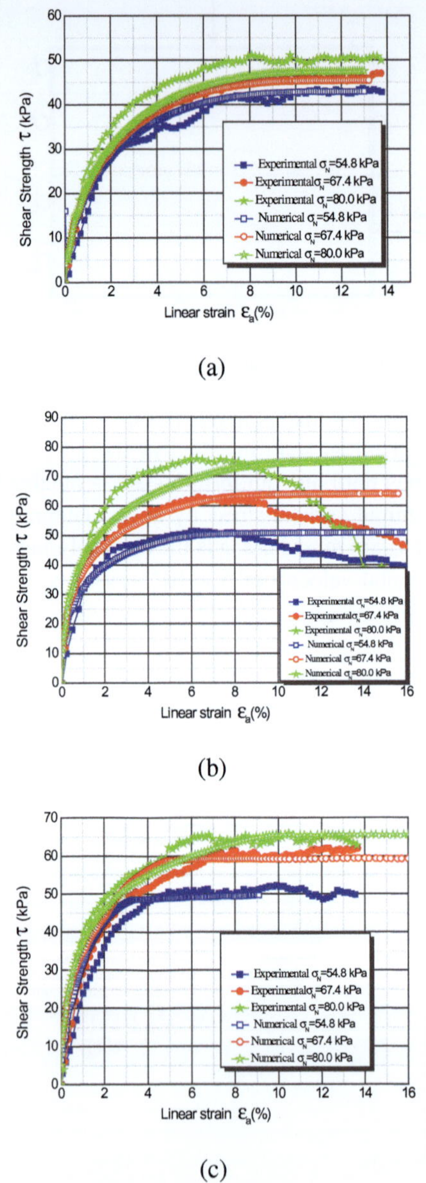

(a)

(b)

(c)

Fig. 11. Comparison of model predictions with experimental results in the case of (a) Clay-Clay, (b) Sand-sand and Clay-Sand

6   Conclusion

The shear behavior of sand/clayey soil interface using a simple direct shear test is studied. In the shear test a shearing surface is produced and the resistance to shearing is defined as the shear resistance. The frictional resistance is, however, determined on an existing joint surface. The frictional behavior at the soil– other materials interface is usually and commonly obtained from direct shear tests.

The built-in interface element in FLAC3D can simulate the soil-soil interfaces, i.e. relationship between shear stress and shear displacement, according to the linear elastic-perfectly plastic model. The used input parameters are derived and calibrated from experimental results. It can be seen that the numerical model is able to reproduce the behavior of soil-soil interface with different normal stresses. Some elaborated models can be also used according to nonlinear strain-softening interface behavior (Wu et al. 2011).

References

Aboshi, H., Ichimoto, E., Harada K., and Enoki, M.: The composer: a method to improve characteristics of soft clays by inclusion of large diameter sand columns. In: Proceedings of the International Symposium on "Reinforcement of Soils", ENPC-LCPC, Paris, pp. 211–216 (1979)

Anubhav, Basudhar P.K.: Modeling of soil–woven geotextile interface behavior from direct shear test results. Geotext. Geomembr. **28**(2010), 403–408 (2010)

Bachus, R.C., Barksdale, R.D.: Vertical and lateral behaviour of model granular columns. In: Proceedings of the International Conference on In-situ Soil and Rock Reinforcement, Paris, pp. 99–110 (1984)

Balaam, N.P., Booker, J.R.: Effect of stone column yield on settlement of rigid foundations in stabilized clay. Int J. Num. Anal. Meth. Geom. **9**(4), 331–351 (1985)

Bergado, D.T., Ramana, G.V., Sia, H.I., Varun: Evaluation of interface shear strength of composite liner system and stability analysis for a landfill lining system in Thailand. Geotext. Geomembr. **24**(6), 371–393 (2006)

Bernal, A., Salgado, R., Swan Jr., R.H., Lovell, C.W.: Interaction between tire shreds, rubber–sand and geosynthetics. Geosynth. Int. **4**(6), 623–643 (1997)

Bouassida, M., Hadhri, T.: Extreme load of soils reinforced by columns: the case of an isolated column. Soils Found. **35**(1), 21–36 (1995)

Bouassida, M., Jellali, B.: Capacité portante d'un sol renforcé par une tranchée. Revue Française de Génie Civil. **6**(7, 8) 1381–1395 (2002)

Datye, K.R.: Settlement and bearing capacity of foundation system with granular columns. In: Proceedings of the Symposium on Soil and Rock Improvement Techniques Including Geotextiles, Reinforced Earth and Modern Piling Methods. AIT–Bangkok, pp. 1–27 (1982). A1

Frikha, W., Tounekti, F., Kaffel, W., Bouassida, M.: Experimental study for the mechanical characterization of Tunis soft soil reinforced by a group of sand columns. Soils Found. (2015). doi:10.1016/j.sandf.2014.12.014

Frikha, W., Bouassida, M.: Prediction of granular column ultimate bearing capacity using expansion cavity model. In: Proceedings of the ICE – Ground Improvement (2015). doi:10. 1680/grim.13.00045

FLAC3D Online Manual. Itasca Consulting Group, Fast Lagrangian Analysis of Continua in 3 Dimensions. Itasca Consulting Group, Minneapolis (2005)

Greenwood, D.A.: Mechanical improvement of soils below ground surface. In: Proceedings of Ground Engineering Conference, pp. 9–20. Institute of Civil Engineering, London (1970)

Hu, W.: Physical Modelling of Group Behavior of Granular Column Foundations. Ph.D. thesis, University of Glasgow (1995)

Hughes, J.M.O., Withers, N.J., Greenwood, D.A.: A field trial of reinforcing effects of stone columns in soil. Géotechnique 25(1), 31–44 (1975)

Lee, K.M., Manjunath, V.R.: Soil–geotextile interface friction by direct shear test. Can. Geotech. J. 37(1), 238–252 (2000)

Liu, C.-N., Ho, Y.-H., Huang, J.-W.: Large scale direct shear tests of soil/PET– yarn geogrid interface. Geotext. Geomembr. 27, 19–30 (2009)

O'Rourke, T.D., Druschel, S.J., Netravali, A.N.: Shear strength characteristics of sand polymer interfaces. J. Geotech. Eng. 116(3), 451–469 (1990). ASCE

Palmeira, E.M.: Discussion on 'Direct shear tests and reinforced sand' by Jewell R.A. and Wroth C.P. Géotechnique 38(1), 146–148 (1988)

Wood, M., Hu, D.W., Nash, D.F.: Group effects in granular column foundations. Géotechnique 50, 689–698 (2000). doi:10.1680/geot.2000.50.6.689

Wu, H.-M., Shu, Y.-M., Zhu, J.-G.: Implementation and verification of interface constitutive model in FLAC3D. Water Sci. Eng. 4(3), 305–316 (2011). doi:10.3882/j.issn.1674-2370.2011.03.007

Xenaki, V.C., Athanasopoulos, G.A.: Experimental investigation of the interaction mechanism at the EPS Geofoam–sand interface by direct shear testing. Geosynth. Int. 8(6), 471–499 (2001)

Effect of the Addition of Chemical Stabilizers on the Characteristics of Clays

Mouloud Benazzoug[1] and Ramdane Bahar[2(✉)]

[1] Geomaterials and Environment Laboratory, University Mouloud Mammeri
of Tizi-Ouzou, Tizi Ouzou, Algeria
[2] Faculty of Civil Engineering, LEEGO, University of Sciences and Technology
Houari Boumediene, Algiers, Algeria
rbahar@usthb.dz

Abstract. An experimental investigation on the shrinkage and durability of stabilized soil was conducted to examine the effect of the addition of chemical stabilizers such as cement, lime, filler and their combination on the characteristics of clays. Shrinkage was measured on un-stabilized soil and stabilized samples with cement, sand or a mixture of cement and sand. The results show that the shrinkage of un-stabilized soil is proportional to the water content of saturation. The higher the percentage of initial water in the soil, the greater the drying shrinkage. For the stabilized soil, the combination of cement and sand reduces the shrinkage slightly better than when only cement is added. The durability was evaluated by conducting series of wetting-drying cycle tests. The study shows that the durability of earth material can be improved by cement and lime stabilizers.

1 Introduction

In the interest of economy, using the locally available materials is one of the main requirements in minimization of construction cost. Faced with an ever increasing problem of providing adequate yet affordable housing in sufficient numbers, the local authorities in Algeria encourage research to develop low cost, readily available and durable building materials. Earth is the oldest building material known, and different techniques are used in earth construction (Stulz and Mukerji 1988; Houben and Guillaud 1989; UNCHS 1987). The oldest technique used is rammed earth or "pisé" which consists of pouring earth stabilized by natural fibres or a binder in a pre-prepared formwork for wall construction with manual compaction in layers of about 1 m height. This technique is suitable for soils with high percentage of large grain size particles. An interesting experience has been reported in North Africa using a similar technique for building low cost grain silos in rural regions using earth stabilised by wheat straws (Bartali 1991). Another technique is the adobe, where blocks are manually prepared in wooden moulds and dried in the open air. Straw is sometimes used to reduce cracking. This technique is mostly used in rural areas and in self-built housing projects.

Earth construction is the most used type of building throughout the long history of Algeria. Some historical cities in the desert and mountainous regions are known of their experience in the use of this low cost material in housing projects. Many failures have

© Springer International Publishing AG 2018
M. Bouassida and M.A. Meguid (eds.), *Ground Improvement and Earth Structures*,
Sustainable Civil Infrastructures, DOI 10.1007/978-3-319-63889-8_10

been reported after seasonal flooding in some cities in Algeria, which undermined the use of earth blocks and led local people to the use of concrete blocks and burned bricks. However, due to its higher cost and lower thermal performance, much interest is going back to earth construction, which is known for its cheap labour and low cost and comparable thermal insulation characteristics. It was revived in early 1980s in Algeria following an instruction from the Algerian Ministry of Housing to local authorities to encourage the use of local construction materials in rural and desert regions. Prototypes and housing projects in different regions were built. Much of the work was on stabilised soil blocks as it is available everywhere and previous historic experience exists.

The use of earth as a building material offered numbers of advantages. However, some of the drawbacks using earth alone for construction are its lack of durability and shrinkage cracking under the action of the weather and its low strength. In addition, with the developments of masonry and reinforced concrete, soil based constructions are regarded as designed for the poor people and hence of lower quality. At present there is a growing market for earth wall buildings moving towards more durable materials of stabilized rammed earth and compressed stabilized earth bricks and blocks. Although the strength is proven to be adequate, there is some consumer resistance for earth building materials considering its durability properties. Much work is needed to identify and enhance durability of earth materials and to convince the general public on durability aspects of earth buildings. Hence, it needs stabilization to increase its strength, durability and shrinkage characteristics. Among the several types of stabilization, the most promising technique is chemical and/or mechanical stabilized soil. A clay sandy soil is usually used after being mixed with some cement or lime in the moulds, hydraulically compacted and then cured. Many research works have been done on soil stabilization using mainly cement and lime stabilization (Gresillon 1978; Houben and Guillaud 1989; Heathcote 1995; Walker 1995; Morel et al. 2001; Bouhicha et al. 2005; Bahar et al. 2004). This paper presents the results of an experimental study on the shrinkage, strength and durability characteristics of local stabilized fine soil. The main objective of this experimental study is to examine the effect of the addition of chemical stabilizers such as cement, lime, filler and their combination on the characteristics of local Algerian clays.

2   Material Characteristics

Three typical soils marl (M), clay (C) and sand (S), from the region of Tizi-Ouzou, east of Algiers, Algeria, were studied. This region is a very populated mountainous region known of its earth construction and local traditional pottery industry. Soils were first passed through a sieve of 5 mm before being characterized for its grading curve, consistency limits and chemical composition. Figure 1 shows the particle size distribution curves of these soils. The sand was fine river sand passing a sieve of 0.63 mm. According to the AASHTO system, it is classified in the group A1. The gradation of the marl (M) and clay (C) are between two boundary curves and relatively close to the

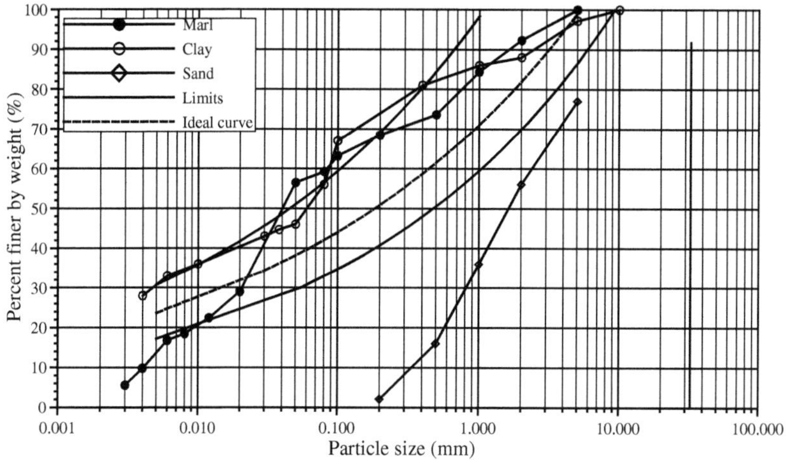

Fig. 1. Grain size distribution

Table 1. Chemical composition of the soils

Sample	Chemical composition (%)										
	SiO_2	Al_2O_3	Fe_2O_3	CaO	MgO	SO_3	K_2O	Na_2O	P_2O_5	TiO_2	PF
Clay (C)	54.46	23.21	7.24	0.44	2.05	0.03	4.29	1.33	0.14	1.19	5.59
Marl (M)	45.40	10.61	5.25	16.14	1.98	0.06	2.05	0.52	0.12	0.57	17.31

ideal curve called Fuller curve (CRATERRE 1991). The marl soil consists of 7.7% gravel, 33.1% sand and 59.2% fine particles and has a plasticity index of 15 and a liquid limit of 39, which ranks in the group A6 fine-grained soils according to the AASHTO system. The liquid limit and plasticity index of the clay soil are 54 and 37 respectively, which ranks in the group A7-6 fine-grained soils according to the AASHTO system. The sand equivalent of the two soils is less than 20. The results of chemical analysis of the tested soils are given in Table 1. The optimum water content for the soil without stabilization was obtained using a normalized Proctor mould. The results are shown in Fig. 2. The clay sample is composed of quartz and a clay fraction (muscovite, kaolinite, feldspar and chlorite). The marl sample is composed mainly of carbonates (limestone, dolomite) of quartz and clay minerals (muscovite, kaolinite, feldspar). The stabilizer materials used are fine fillers added with a grain size varying from 0 to 125 µm, Ordinary Portland cement (CEMII 32.5) and lime obtained by firing limestone at a temperature of 1000 °C.

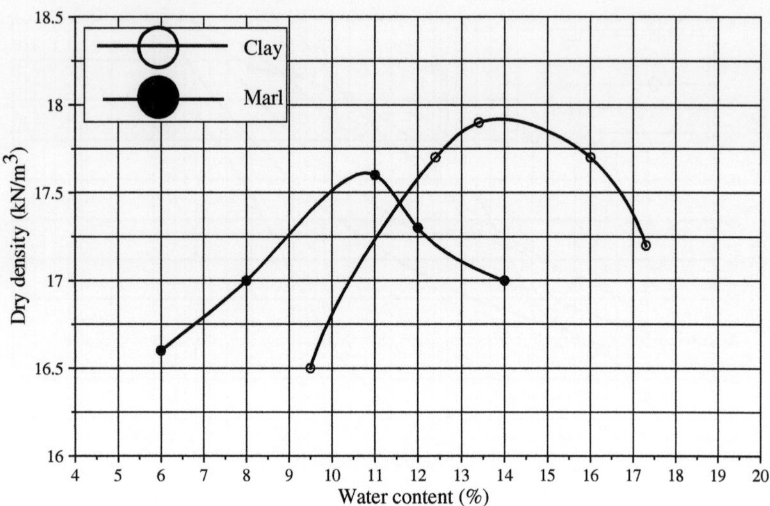

Fig. 2. Normalized Proctor curve.

3   Testing Program

Unstabilized and stabilized samples for testing are prepared by first drying the soil in an oven, and a homogeneous mixture obtained by blending the required amount of cement with the dry soil in a mechanical mixer before adding water and further mixing. The soil mixture is put in a normalized Proctor mould and is compacted dynamically. Kenai et al. (2006) show that mechanical stabilization by dynamic compaction of local clay sandy soil seems to enhance the mechanical properties and water resistance of the soil as compared to the static or the vibro-static compaction methods. The tests were conducted according to the local and to the RILEM TC 153 recommendations (CNERIB 1993, 1994; RILEM 1995).

3.1   Shrinkage Tests

To test the influence of the water content on the soil shrinkage, we conducted soil shrinkage measures with the measuring device described in the scheme of Fig. 3. The test consists of daily measure the shortening of the length of the sample on the comparator. We will also weigh these specimens on a scale; the test is continued until stabilization of the mass. After heating in the oven at 105 °C for 24 h, the specimen is then measured (dry length L_f) and weighed (M_s). The test lasts about twenty days.

3.1.1   Unstabilized Soil

Three water content levels have been considered for the samples: 10%, 14% and 30% for samples E10, E14 and E30 respectively. These initial water contents are chosen in relation to the natural water contents measured on the marls and clays of the Tizi-Ouzou region in summer and winter. Figure 4 illustrates the variation of the length

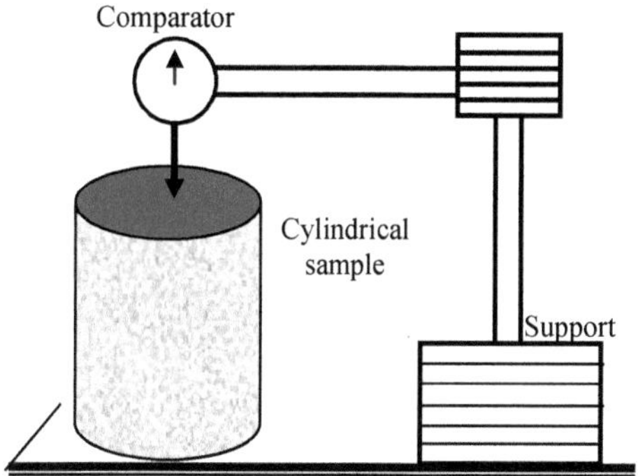

Fig. 3. Experimental measurement of shrinkage

of the sample as a function of the age of the specimen. Figure 4 shows a reduction in the sample length during the first five days of drying, followed by a tendency to stabilize. This decrease is more pronounced for high initial water contents. For the E10 sample, the length does not vary from the 10th day. For E14 and E30 sample, the length does not vary from the 7th day and the 11th day respectively. It is noted that most the initial water content is high, the greater the decrease in length is important. Large variations in length are observed during the first days after the making of the specimens. The amplitude of the shortening of the length of the samples is proportional to the initial water content. When the sample dries, the water volume decreases. The samples change volume and the curves of variation of the length as a function of time have two phases. In the first one, the length variation is significant during the first 5 days, the grains are close to each other, the sample is always saturated and the length of

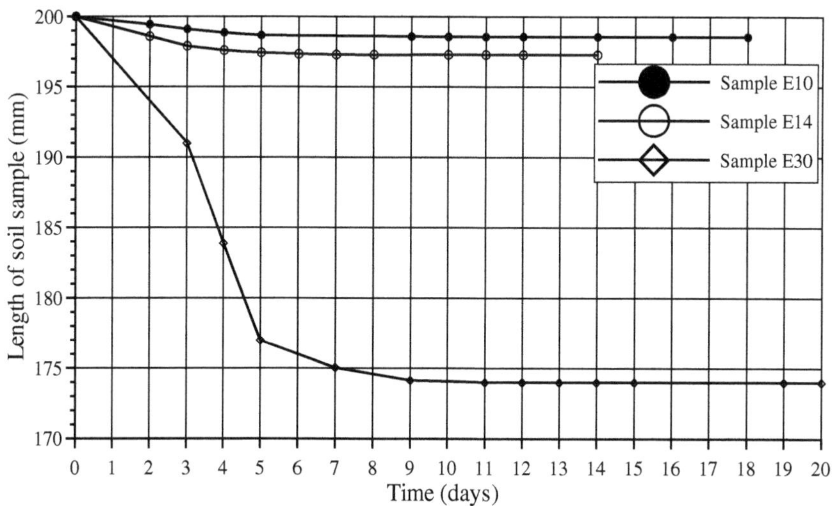

Fig. 4. Evolution of the sample length as a function of time for the three samples

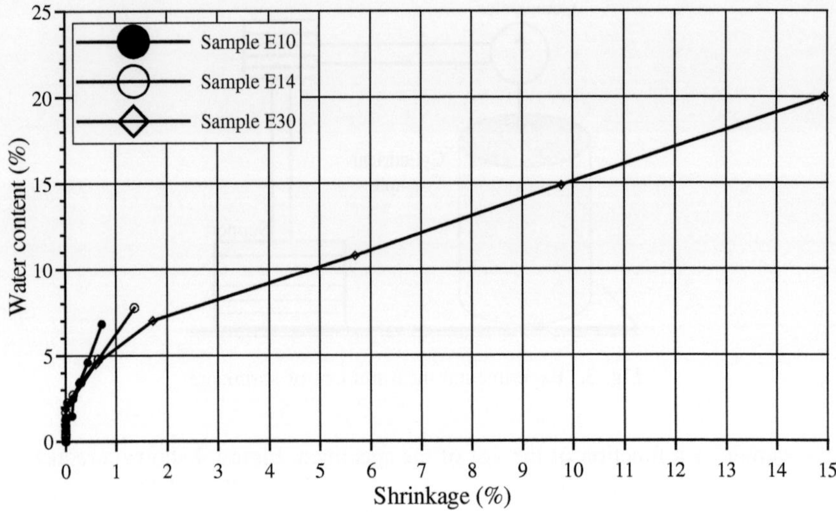

Fig. 5. Evolution of water content as a function of shrinkage (E10, E14, E30)

Table 2. Shrinkage limits

Sample	Effective shrinkage limit w_{re} (%)
E 10	1.47
E 14	2.26
E 30	5

the specimen decreases significantly. In the second one, the grains are in contact and can no longer close. The shortening is slowed and eventually stabilizes at the end of the 10th day. Figure 5 clearly shows the two main phases of shrinkage: normal shrinkage and absence of shrinkage. The normal shrinkage is proportional to the water content of saturation. The higher the percentage of initial water contained in the soil is high, over the drying shrinkage is important. The second phase is characterized by shrinkage equal to zero; the removal of water does not vary the length of the sample. The intersection of the line representing normal shrinkage with the axis carrying the moisture (absence of shrinkage) gives the effective shrinkage limit (Table 2).

3.1.2   Stabilized Soil

Shrinkage was measured on samples stabilized with cement, sand or a mixture of cement and sand. Specimens were cured inside the laboratory at about 25 °C and 65% R.H. Figure 6 shows typical variation of shrinkage with time for cement-stabilized soil. Shrinkage increases rapidly during the first four days for both cement stabilized soil and un-stabilized soil specimen and then at a later ages the increase is very slow. Hence, curing for the first four days could be beneficial in reducing drying shrinkage and cracking. The shrinkage of cement stabilized soil at 25 days of age as compared to that of un-stabilized soil was reduced by about 20% and 44% for 6% and 10% of

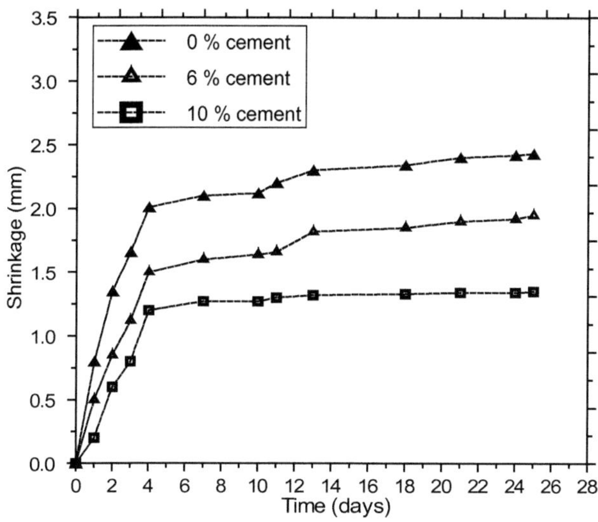

Fig. 6. Effect of cement content on the development of shrinkage.

cement content respectively. A study was also conducted on the effect of using a combination of sand and cement on the final shrinkage at 25 days of age (Table 3). As expected, the addition of sand reduces the shrinkage as sand particles oppose the shrinkage movement. The reduction in shrinkage was about 29% and 64% for 10% and 15% of sand content respectively. Further increase of sand content did not affect the shrinkage. The combination of cement and sand reduces the shrinkage slightly better than when only cement is added. When a combination of cement and sand was added, a mixture of 5% sand and 15% of cement seems to give the lowest shrinkage.

Table 3. Variation of final shrinkage with water, sand and cement contents

Water content (%)	Final shrinkage (mm)	Sand content (%)	Final shrinkage (mm)	Cement (C) + Sand (S)	Final shrinkage (mm)
8.63	0	5	1.74	5%C + 15%S	1.12
10.0	0	10	1.73	10%C + 10%S	1.11
12.0	0.86	15	0.88	15%C + 5%S	0.74
13.43	2.59	20	0.87		
15.56	3.80				
19.28	6.03				

3.2   Wetting - Drying Cycle Tests

The purpose of the durability test is to evaluate the ability of the stabilized material to resist to the water action. This ability is often studied by wetting-drying cycle test, weight loss test and absorption test by immersion and capillary action. In this paper, only the results obtained by tests of wetting–drying cycle tests carried out on stabilized clay samples are presented. Compressive strength was determined on samples prepared in compaction moulds under standard Proctor conditions. Tests were conducted according to the local (CNERIB 1993, 1994) and to the RILEM TC 153 (RILEM 1995) recommendations. Samples for testing are prepared by first drying the soil in an oven and a homogeneous mixture obtained by blending the required amount of cement, lime, filler or combination of these three stabilizers with the dry soil in a mechanical mixer before adding water and further mixing.

The following nomenclature was adopted for additions and their contents used in the preparation of samples.

C: Cement	CY: Y% cement content (C5 = 5%, C10 = 10%, C15 = 15%)
F: Filler	FY: Y% filler content (F2 = 2%, F5 = 5%, F10 = 10%)
X: Lime	XY: Y% lime content (X2 = 2%, X5 = 5%)
S: Sand	SY: Y% sand content (S20 = 20%, S30 = 30%)

In order to study the durability of stabilized soil material when subjected to alternating precipitation (water saturation in winter, dry in summer), a standard test of sustainability (wetting - drying) was performed. The samples are subjected to five cycles of wetting - drying. They are firstly immersed for 5 h in water and then dried in an oven for 24 h at a temperature of 105 °C. In the last cycle, they are subjected to the compression test immediately after immersion. The compressive strength of clay stabilized with cement, lime, filler, addition of sand and their combinations after five cycles of wetting-drying is given in Figs. 7, 8 and 9. Observations made during these tests show that the untreated samples were completely destroyed in the early cycles, indicating the low water resistance of the used materials. Stabilized specimens show a better resistance to water.

For the clay stabilized only by cement, Fig. 7 shows that the compressive strength, after 5 cycles of wetting-drying, increases with the increase cement content. An addition of 10% of cement increases 10 times the resistance to water of the stabilized material with 5% cement content. The addition of 15% of cement causes an increase in resistance smaller than that of the material stabilized with 10% cement. These results indicate that the resistance of the clay stabilized in the range 5 to 10% is very sensitive to cement content. Strength gain is relatively low beyond 10% cement content. If we consider the strength limit of 4 MPa imposed by standards for low rise buildings (less than 2 floors) the addition of 10% cement provides a satisfactory safety factor of 3.4. For clay stabilized by mix of cement and lime, Fig. 7a clearly shows that the strength is high for higher cement content. Figure 7b shows that the curves for 2% and 5% of lime are located above the curve for treatment with cement alone, and intersect in a point corresponding to a cement content of about 13%. These results clearly indicate that the lime content depends on the effective cement content. For cement contents below 13%, the best strength is obtained for a lime content of 5%. Beyond 13% of cement, 2% lime gives better results. For an addition of 10% cement and 5% lime, strength gain is about 92%, almost double the strength obtained by treatment with cement alone. The increase in strength was 84% for 2% lime combined with a cement content of 15%. The best performance of the material was obtained with 15% of cement and 2% lime

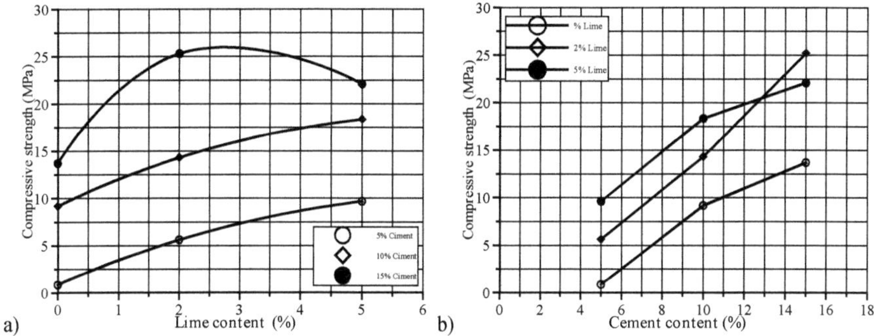

Fig. 7. Lime stabilization: compressive strength after 5 cycles wetting - drying (a) as a function of lime content, (b) as function of cement content.

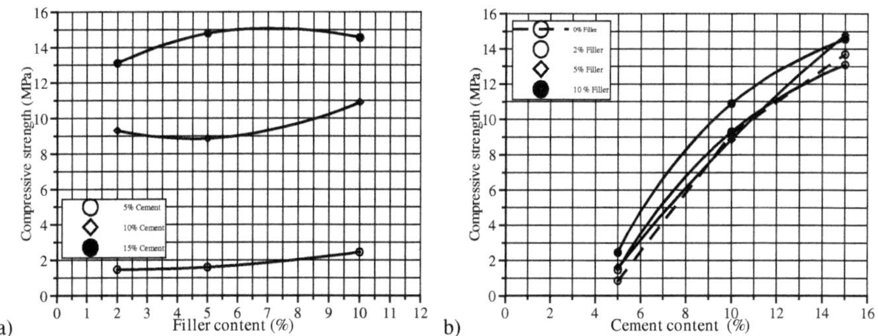

Fig. 8. Mix (cement – filler) stabilization: compressive strength after 5 cycles wetting – drying, (a) as a function of filler content, (b) as function of cement content.

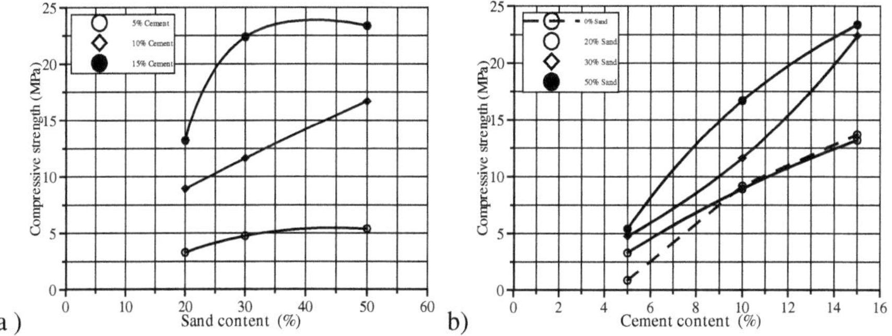

Fig. 9. Mix cement – sand stabilization: compressive strength after 5 cycles wetting - drying (a) as a function of sand content, (b) as function of cement content.

(25.2 MPa). For clay stabilized by a combination of cement and filler, Figs. 8a and b show that the addition of 2 and 5% filler generate an insignificant strength gain and may even bring down the strength of the material stabilized with cement alone. The addition of 10% filler slightly improves the strength. Strength gain as compared to treatment by cement alone is 6.2% and 8.7% for (15% cement and 10% fillers) and (10% cement and 10% fillers), respectively. This slight improvement is probably due to the improvement of the compactness of the material induced by the reduction of voids between the integration of fine soil particles. The addition of fillers is interesting for a content of 10% which causes a slight strength gain whatever the cement content.

Figure 9 shows the influence of the addition of sand on the compressive strength of clay stabilized by mix of cement and sand. The incorporation of cement has a positive effect on the strength, the higher the cement content, the higher is the strength. The best results are obtained with 15% cement and 50% sand (23.4 MPa). Figure 9 shows the same results as gains, losses and performance compared to those obtained for the cement stabilized with clay alone. Figure 9a shows that the strength gain is very important for low cement contents (5%). It is observed that if the cement content is maintained at 5% for a particular application of stabilization, the incorporation of sand is extremely useful because an addition of 20% sand increase by 2.8 time the strength of the cement stabilized clay alone and an addition of 30% and 50% of sand, this improvement is 4.5 and 5.2 times larger respectively. The best strength gains are given by the cement content of 5%. For 10 and 15% cement content, Fig. 9b shows that gains in strength are quite similar and the curves obtained are almost identical.

Figure 10 shows the influence of the addition of a combination of cement, filler and sand on the compressive strength of stabilized clay. For clay stabilized with 20% sand, the addition of 2% lime seems to give interesting results for cement contents ranging only between 8 and 11%. Outside this range, the strength is lower than that obtained with the addition of 5% of lime. With 5% lime, the strength varies linearly with the cement content. However, in the range of 8 to 11% of cement, it was found that the

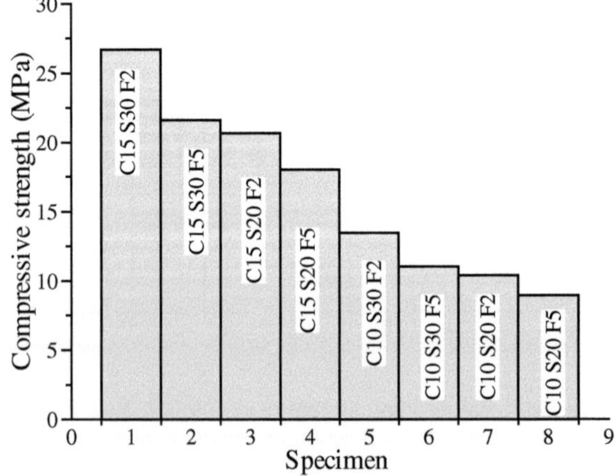

Fig. 10. Compressive strength, stabilized clay with mix (cement-sand–filler).

treatment with 5% lime causes a loss of strength of 3%, which is very low. It can be concluded that treatment with 20% sand can produce quite good results with lime content of 5%. For clay stabilized with 30% sand, there is a similar situation than with 20% sand, but with a higher effect on the strength for 2% lime. The addition of 2% lime can be interesting for cement content between 8 and 14%. Stabilization with incorporation of 5% lime significantly improves resistance to water after 5 cycles wetting - drying. The best performance is achieved with (15% Cement + 30% Sand + 2% Lime), where a maximum strength of 30.6 MPa is obtained.

4   Conclusions

The effect of the addition of chemical stabilizers and their combination on the shrinkage, strength and durability characteristics of clays was examined. The results show that the shrinkage of un-stabilized soil is proportional to the water content of saturation. The higher the percentage of initial water contained in the soil is high, over the drying shrinkage is important. For the stabilized soil, the combination of cement and sand reduces the shrinkage slightly better than when only cement is added. The wetting-drying cycle tests show that the durability of earth material can be improved by the addition of cement and lime. The results show that although adding more than 10% cement improves dry compressive strength. Even with high levels of cement (15%), the compressive strength is altered and is only 88% of the dry strength after cycles of wetting-drying. For low levels of cement content, sand addition has a great influence on strength; the gain obtained by sand incorporation into the clay is about 5 times larger, especially in the presence of high levels of cement. Clay stabilized with a combination of (cement + sand + lime) gives excellent results than the stabilizer by a combination (cement + sand + filler).

References

Bahar, R., Benazzoug, M., Kenai, S.: Performance of compacted cement stabilised soil. Cement Concr. Compos. **25**(5), 633–641 (2004)

Bartali, H.: La terre armée de paille pour les constructions. In: Proceedings of the French-Maghreb Symposium on Construction with Local Materials, Marseille, France, p. 551 (1991)

Bouhicha, M., Aouissi, F., Kenai, S.: Performance of composite soil reinforced with barely straw. Cement Concr. Compos. Special Theme Issue, Natural Fibre Reinforced Cement Composites **27**(9–10), 617–621 (2005)

CNERIB: Recommandations pour la production et mise en œuvre du béton de terre stabilisée. CNERIB, Algiers, Algeria (1993)

CNERIB: Guide technique du béton de terre stabilisée. CNERIB, Ministry of Housing, Algiers, Algeria (1994)

CRATerre: The Basics of Compressed Earth Blocks. A publication of the Deutsches Zentrum fur Entwicklungtechnologien-Gate (1991)

Gresillon, J.M.: Etude de l'aptitude des sols à la stabilisation au ciment application à la construction. Annales de l'ITBTP, n° 361, Paris, pp. 2–8 (1978)

Heathcote, K.A.: Durability of earthwall buildings. Constr. Build. Mater. 9(3), 185–189 (1995)

Houben, H., Guillaud, H.: Traité de construire en terre: l'encyclopédie de la construction en terre, CRATerre, vol. 1, Edition Parenthèse, Paris (1989)

UNCHS: Earth construction technology, Technical notes No. 11 (1987)

Kenai, S., Bahar, R., Benazzoug, M.: Experimental analysis of the effect of some compaction methods on mechanical properties and durability of cement stabilized soil. J. Mater. Sci. 41(21), 6956–6964 (2006)

Morel, J.C., Mesbah, A., Oggero, M., Walker, P.: Building houses with local material: means to drastically reduce the environmental impact of construction. Build. Environ. 36, 1119–1126 (2001)

RILEM TC 153 - CIB W90 CEB: Technologie du bloc de terre comprimée. Modes opératoires pour les essais d'identification en laboratoire des terres (1995)

Stulz, R., Mukerji, K.: Appropriate Building Materials – A Catalogue of Potential Solutions. SKAT & IT Publications, Bern (1988)

Walker, P.J.: Strength, durability and shrinkage characteristics of cement stabilized soil blocks. Cement Concr. Compos. 17(4), 301–310 (1995)

Boundary Element Analysis of Shear-Deformable Plates on Tension-Less Winkler Foundation

Ahmed Fady Farid[1] and Youssef F. Rashed[1,2(✉)]

[1] Department of Structural Engineering, Cairo University, Giza, Egypt
[2] Supreme Council of Universities, Giza, Egypt
yrashed@scu.eg

Abstract. A new direct Boundary Element (BEM) technique is established to analyze plates on tension-less elastic foundation. The soil is modeled as Winkler springs. The considered BEM is based on the formulation of shear deformable plate bending theory according to Reissner. The developed technique is based on coupling the PLPAK software with iterative process to eliminate tensile stresses underneath the considered plate. Tensile zones are redistributed until the final contact zone of plate is reached. Examples are tested and results are compared to analytical and previously published results to verify the proposed technique.

1 Introduction

Solving plates on tensionless foundation can be divided into two main categories of solution. First solution is to solve the problem using iterative procedure to consider the miscontact between plate and foundation. Second solution is to embedded the contact problem in a system of nonlinear equations and solve it using optimization algorithm. Boundary element method (BEM) is widely used to solve plate on foundation problem. Katsikadelis and Armenakas [1] presented analysis of thin plates on elastic foundation. A BEM formulation based on shear deformable plates according to Reissner [2] was derived by Vander Weeën [3]. Rashed et al. [4–6] extended Vander Weeën formulation to model foundation plates.

Several studies are presented to solve plates on tensionless foundations. Weitsman [7] presented analysis of tensionless beams, or plates, and their supporting Winkler or Reissner subgrade due to concentrated loads. Celep [8] presented the behavior of elastic plates of rectangular shape on a tensionless Winkler foundation using auxiliary function. Galerkin's method is used to reduce the problem to a system of algebraic equations. Li and Dempsey [9] used an iterative procedure to analyze unbonded contact of a square thin plate under centrally symmetric vertical loading on elastic Winkler or EHS foundation. Sapountzakis and Katsikadelis [10] presented boundary element solution for unilateral contact problems of thin elastic plates resting on linear or

© Springer International Publishing AG 2018

M. Bouassida and M.A. Meguid (eds.), *Ground Improvement and Earth Structures*, Sustainable Civil Infrastructures, DOI 10.1007/978-3-319-63889-8_11

nonlinear subgrade by solving a system of nonlinear algebraic equations. Xiao et al. [11] presented a BE-LCEM solution for thin free edge plates on elastic half space with unilateral contact. Silva et al. [12] used finite element method to discretize the plate and foundation then used three alternative optimization linear complementary problems to solve plates on tensionless elastic foundations. T-element analysis of plates on uni-lateral elastic Winkler type foundation using hybrid-Trefftz finite element algorithm was presented by Jirousek [13]. Xiao [14] presented a BE-LCEM solution to solve unilateral free edges thick plates. Nonlinear bending behavior of Reissner-Mindlin plates with free edges resting on tensionless elastic foundations using admissible functions was presented by Hui-Shen and Yu [15]. Silveira et al. [16] presented a nonlinear analysis of structural elements under unilateral contact constrains studied by a Ritz approach using a mathematical programming technique. Results of finite element analysis of beam elements on unilateral elastic foundation using special zero thickness element designed for foundation modeling is presented by Torbacki [17]. Buczkowski and Torbacki [18] presented finite element analysis of plate on layered tensionless foundation. Kongtng and Sukawat [19] used the method of finite Hankel integral transform techniques for solving the mixed boundary value problem of unilaterally supported rectangular plates loaded by uniformly distributed load.

In this paper, an iterative procedure is developed to solve thick plates on tensionless Winkler foundation. The boundary element formulation is used to extract the stiffness matrix of the plate supported on Winkler springs. The main advantage of the proposed technique that it merge between the advantage of the boundary element method, modeling plate using integral equation, and the simplicity of finite element method, solving nonlinear equations iteratively in matrix form. One of the advantages of the proposed technique is its practicality. In the formulation of Xiao [19] although solving of thick plates on tensionless foundations, the results were not accurate near corners as it will be shown in the examples of this paper. Also, this formulation is suitable to be extended for solving plates on Winkler elastic-plastic foundations. Numerical examples are presented to verify efficiency and practicality of the proposed technique.

2   BEM for Plate on Winkler Foundation

Formulation of direct boundary integral equation of the plate [6] based on Reissner plate bending theory [2] is used. Consider a general plate domain Ω and boundary Γ with internal Winkler cells as shown in Fig. 1. The indecial notation is used in this paper where the Greek indexes vary from 1 to 2 and Roman indexes vary from 1 to 3. The integral equation can be represented as follows:

$$C_{ij}(\xi)u_j(\xi) + \int_{\Gamma(x)} T_{ij}(\xi, x)u_j(x)\,d\Gamma(x) = \int_{\Gamma(x)} U_{ij}(\xi, x)t_j(x)\,d\Gamma(x)$$

$$+ \sum_c \int_{\Omega(y)} \left[U_{ik}(\xi, y) - \frac{v}{(1-v)\lambda^2} U_{i\alpha,\alpha}(\xi, y)\delta_{3k} \right] F_k(y)\,d\Omega(y) \tag{1}$$

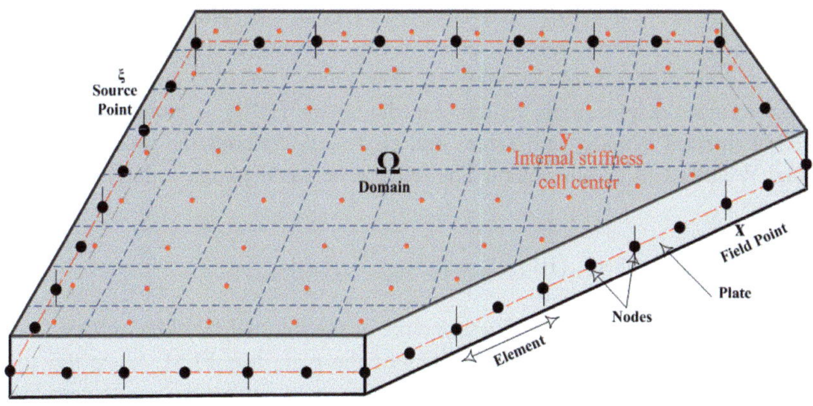

Fig. 1. General plate for BEM formulation of the plate.

Where $T_{ij}(\xi,x)$, $U_{ij}(\xi,x)$ are the two-point fundamental solution kernels for tractions and displacements respectively [3]. The two points ξ and x are the source and the field points respectively. $u_j(x)$ and $t_j(x)$ denote the boundary generalized displacements and tractions. $C_{ij}(\xi)$ is the jump term. The symbols ν and λ denote the plate Poison's ratio and shear factor. c denotes the number of internal Winkler cells that having domain Ω, F_k denotes the Winkler cell two bending moments ($F_1 = M_{xx}$, $F_2 = M_{yy}$) and column vertical force ($F_3 = F$). The field point y denotes the point of the internal Winkler cell center.

After discretizing the boundary of the plate to NE quadratic elements each node has three unknowns. Due to internal Winkler cells inside the domain additional three unknowns are added, so another collocation scheme is carried out at each internal Winkler cell center to add additional equations. The collocation scheme is carried out as follows:

$$u_i(Y) + \int_{\Gamma(x)} T_{ij}(Y,x)u_j(x)\,d\Gamma(x) = \int_{\Gamma(x)} U_{ij}(Y,x)t_j(x)\,d\Gamma(x)$$

$$+ \sum_c \int_{\Omega(y)} \left[U_{ik}(Y,y) - \frac{\nu}{(1-\nu)\lambda^2}U_{i\alpha,\alpha}(Y,y)\delta_{3k}\right]F_k(y)\,d\Omega(y) \tag{2}$$

Where Y is a new source point located at each cell center.

Equations (1) and (2) after discretization of plate boundary can be re-written in a matrix form as follows:

$$\begin{bmatrix} [A]_{3Nx3N} & [A_2]_{3Nx3Nc} & [0]_{3Nx3Nc} \\ [A_1]_{3Ncx3N} & [A_3]_{3Ncx3Nc} & [I]_{3Ncx3Nc} \end{bmatrix} \begin{bmatrix} \{u/t\}_{3Nx1} \\ \{F\}_{3Ncx1} \\ \{u_c\}_{3Ncx1} \end{bmatrix} = \begin{bmatrix} \{0\}_{3Nx1} \\ \{RHS\}_{3Ncx1} \end{bmatrix} \tag{3}$$

Where $[A]_{3N\times 3N}$, $[A_1]_{3Nc\times 3N}$, $[A_2]_{3N\times 3Nc}$, $[A_3]_{3Nc\times 3Nc}$ and $\{RHS\}_{3Nc\times 1}$ contains the coefficients of the integrals presented in Eqs. (1) and (2). $[0]_{3N\times 3Nc}$, $[I]_{3Nc\times 3Nc}$ are the null and identity matrices respectively. The vector $\{u/t\}_{3N\times 1}$ contains the unknown boundary values displacement or traction, the vector $\{F\}_{3Nc\times 1}$ contains the unknown values of internal Winkler cell forces and $\{u_c\}_{3Nc\times 1}$ contains the unknown values of internal Winkler cell displacement. The relation between $\{u_{3Nc\times 1}$ and $\{F\}_{3Nc\times 1}$ can be as follows:

$$\{F\}_{3Nc\times 1} = [K^c]_{3Nc\times 3Nc}\{u_c\}_{3Nc\times 1} \tag{4}$$

Where $[K^c]_{3Nc\times 3Nc}$ is the stiffness matrix of the internal Winkler cells.

In order to extract the stiffness matrix of the plate like FEM. Since the stiffness matrix of the plate is independent on the loading, then neither the no domain load nor the cells area loading is considered. As a result of this assumption the vector $\{RHS\}_{3Nc\times 1}$ will be zeros consequently, Eq. (3) can be re-written as follows:

$$\begin{bmatrix} [A]_{3Nx3N} & [A_2]_{3Nx3Nc} \\ [A_1]_{3Ncx3N} & [A_3]_{3Ncx3Nc} \end{bmatrix} \left\{ \begin{array}{c} \{u/t\}_{3Nx1} \\ \{F_c\}_{3Ncx1} \end{array} \right\} = \left\{ \begin{array}{c} \{0\}_{3Nx1} \\ -\{u_c\}_{3Ncx1} \end{array} \right\} \tag{5}$$

In order to get $\{F_c\}_{3Ncx1}$ represents a certain case of the stiffness $\{u_c\}_{3Ncx1}$ is forced to be unity. Cases of loading equal to the $3N_c$ are considered to extract the plate stiffness matrix as follows:

$$\begin{bmatrix} [A]_{3Nx3N} & [A_2]_{3Nx3Nc} \\ [A_1]_{3Ncx3N} & [A_3]_{3Ncx3Nc} \end{bmatrix} \left\{ \begin{array}{c} \{u/t\}_{3Nx3Nc} \\ \{K_p\}_{3Ncx3Nc} \end{array} \right\} = \left\{ \begin{array}{c} \{0\}_{3Nx3Nc} \\ -\{I\}_{3Ncx3Nc} \end{array} \right\} \tag{6}$$

Where $\{K_p\}_{3Ncx3Nc}$ is the required plate stiffness matrix.

Winkler stiffness values of each cell are added as a diagonal matrix to the plate stiffness matrix $[K_p]$ to extract $[K_{Overall}]_{3Nc\times 3Nc}$ the overall stiffness of the plate rested on elastic foundation. Also condensed load vector of plate is computed $\{P_{plate}\}_{3Nc\times 1}$. The overall equilibrium equation can be written as follows:

$$[K_{Overall}]_{3Nc\times 3Nc} * \{u_{plate}\}_{3Nc\times 1} = \{P_{plate}\}_{3Nc\times 1} \tag{7}$$

3 Proposed Iterative Procedure

In this section, simulation of soil as tensionless material is established using a new iterative procedure to solve plate on tensionless foundation. Figure 2 demonstrates flowchart of the iterative procedure. The system of equations in Eq. (7) is solved then the internal forces of soil are computed. The internal forces of soil are used to get the most tensile stressed cell. Elimination procedure is used to eliminate the most tensile stressed cell DOF stiffness matrix to simulate the loss of contact between plate and soil. The system of equations is resolved. This iterative procedure is used until reaching the real contact zone i.e. no tensile stresses appear in the soil reactions. In order to optimize the solution the formulation is implemented into a computer code. It is worth to be noted that this procedure is valid also in solving plates on elastic perfectly plastic soil by applying a virtual load equal to the allowable compressive stress of the soil at the eliminated cells DOFs.

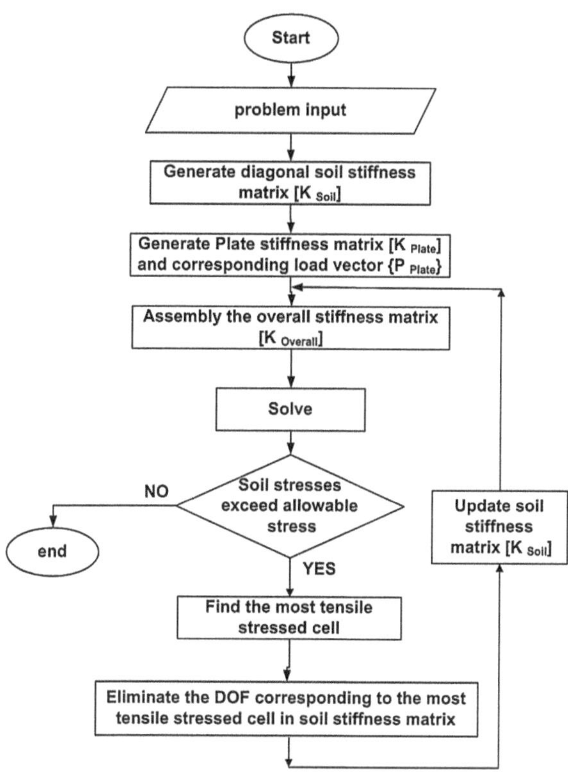

Fig. 2. Flowchart for developed iterative procedure.

4   Numerical Examples

Example 1: Simply supported beam on tensionless Winkler foundation.

Consider the simply supported beam having the width sides of the plate restrained by one pinned element/side as shown in Fig. 3. The beam is subjected to concentrated moments $M = 100$ along its edges as shown in Fig. 3, Poisson's ratio of the plate material is $v = 0$, foundation Winkler stiffness parameter K is 71.68.

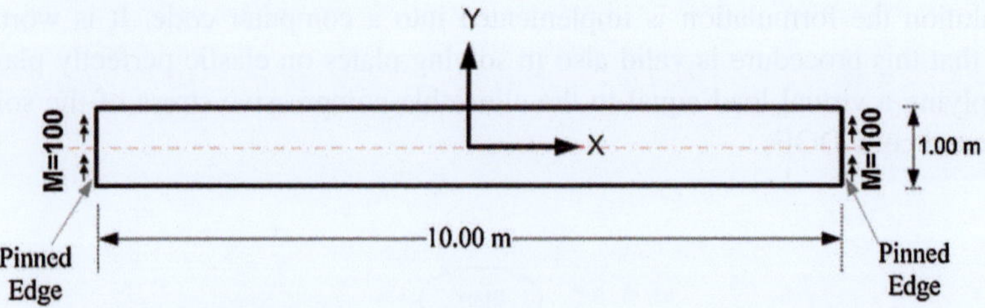

Fig. 3. Simply supported beam resting on Winkler foundation and subjected to concentrated moments (M = 100) on its edges.

The simply supported beam is modeled as thick plate with plate dimensions a = 10, b = 1.0 and t = 0.4, the Young's modulus is E = 10^6. The soil is modeled as Winkler with 40 cells as shown in Fig. 4. It has to be noted that any unit system is possible.

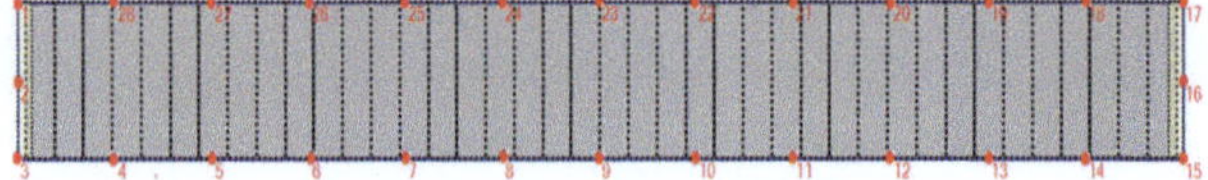

Fig. 4. Beam proposed technique model on Winkler foundation.

The beam is solved by the proposed technique and results are compared to results obtained from Silva et al. [12]. It can be seen from results that the proposed technique of obtained the same contact length (x/a = 0.43) which was also obtained in [12]. In addition, deflection along centerline of the beam (x-axis) and contact reaction are in a good agreement with results of ref. [12] as shown in Figs. 5 and 6.

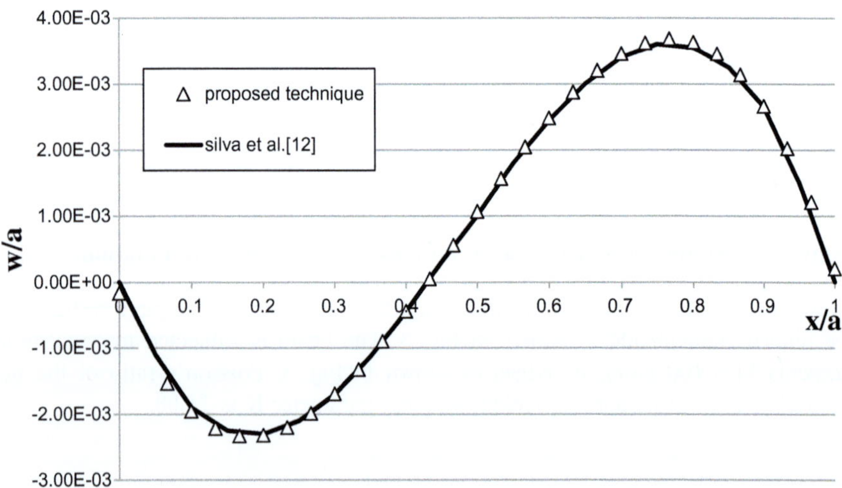

Fig. 5. Deflection along the beam centerline in Example 1.

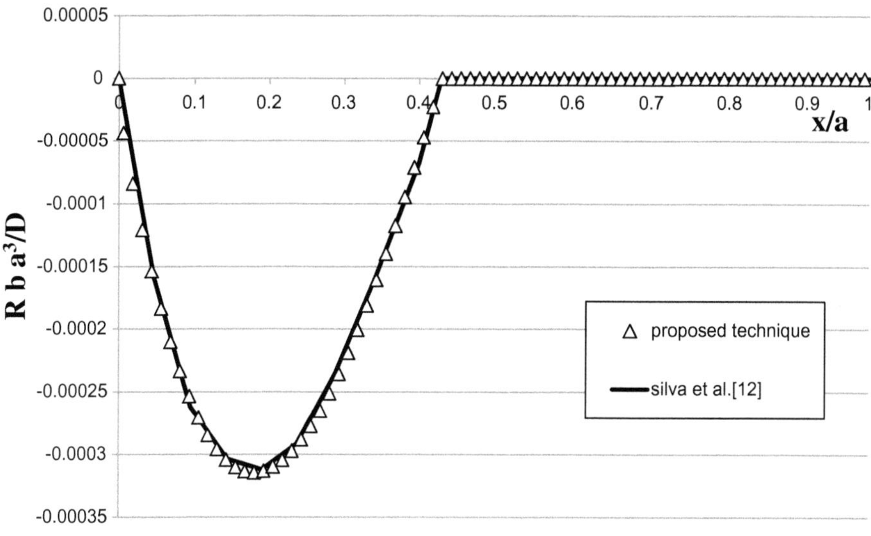

Fig. 6. Contact reaction along the beam centerline in Example 1.

Example 2: Free edge square plate on tensionless Winkler foundation.

This example consists of free edges square plate. The side of plate is a = 400 cm. The Young's modulus is $E = 2.6 \times 10^6$ N/cm^2, Poisson's ratio of the plate material is ($v = 0.15$). Different plate thicknesses (t) are employed 30,40,60,80,100 cm.

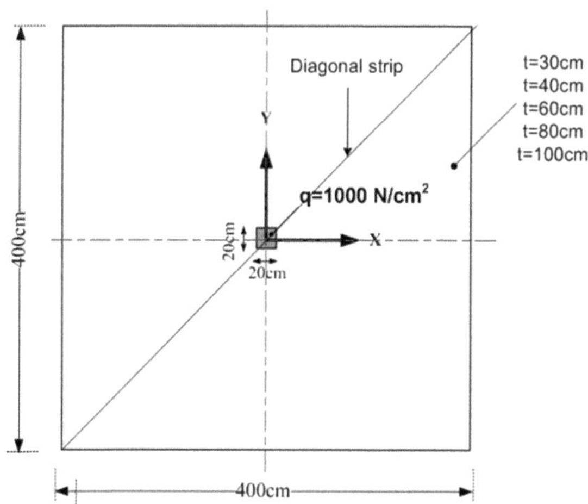

Fig. 7. Free edge square plate dimensions under patch load in Example 2.

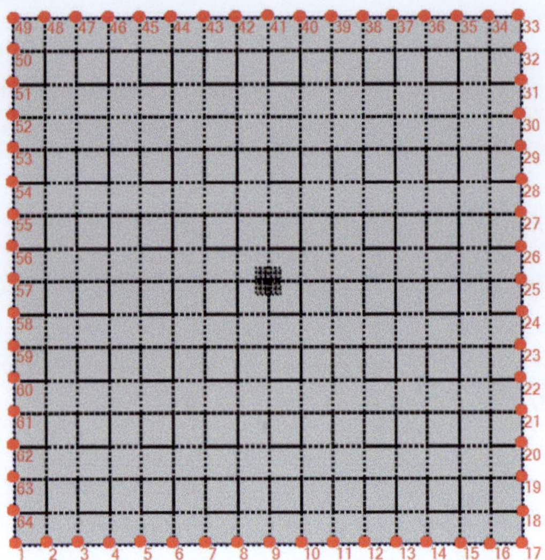

Fig. 8. Proposed technique model of Example 2.

The foundation stiffness parameter K is 500 N/cm^3. Plate is loaded by patch distributed load q = 1000 N/cm^2 over area of 20 cm × 20 cm at its center, as shown in Fig. 7. The plate is modeled as thick plate rested on Winkler foundation as shown in Fig. 7 with soil discretization as shown in Fig. 8.

Results are compared to results obtained in Xiao [14]. All models are solved with different thicknesses (30, 40, 60, 80,100 cm). Table 1 demonstrates the contact regions for different thicknesses of raft. Figures 9, 10, 11 and 12 demonstrate the deflection along diagonal strip (indicated in Fig. 7) for different thicknesses of raft, Xiao [14] results and FEM for both bilateral results (at left of graph) and unilateral results (at right of graph) are plotted also on these figures for the sake of comparison.

It can be seen from results that proposed technique obtained the same contact area that is obtained in [14] and in FEM. Deflection is in a good agreement with results of Xiao [14] except at corners, at which some inaccurate results due to singularity problems near boundary elements appear in the formulation of Xiao [14]. It can be seen that the presented formulation results are in good agreement with results from FEM. This demonstrates the advantages of the proposed technique over the formulation of Xiao [14].

Table 1. Contact region for different thicknesses in Example 2.

t (cm)	Proposed technique results	FEM results
30	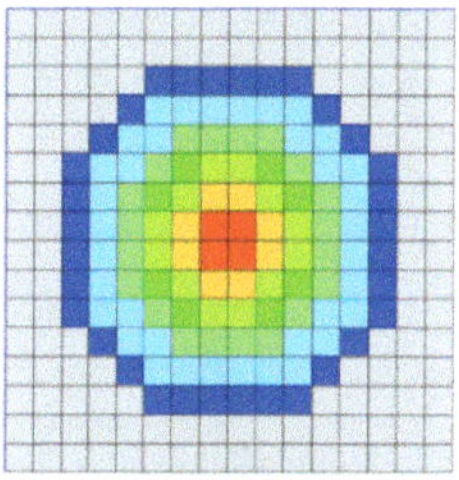	
40	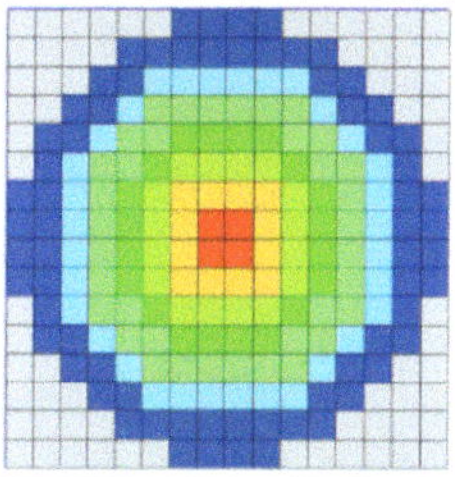	
60	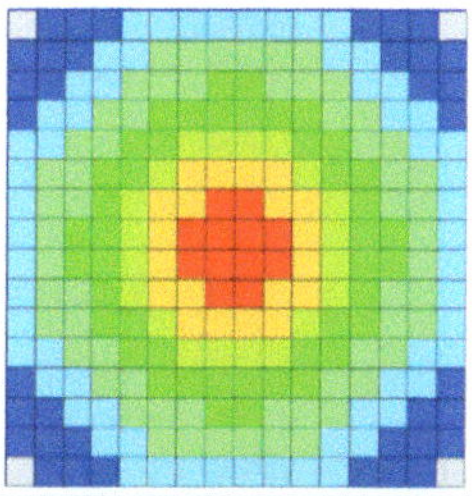	
80		
100	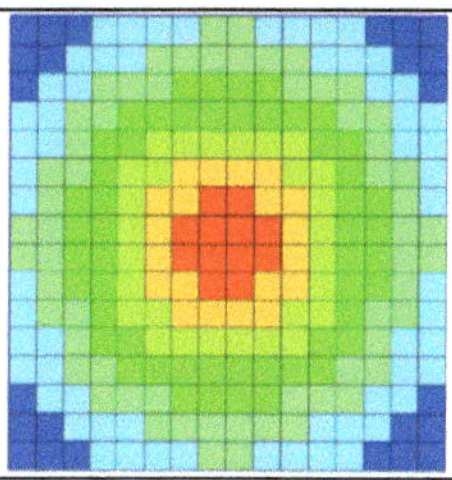	

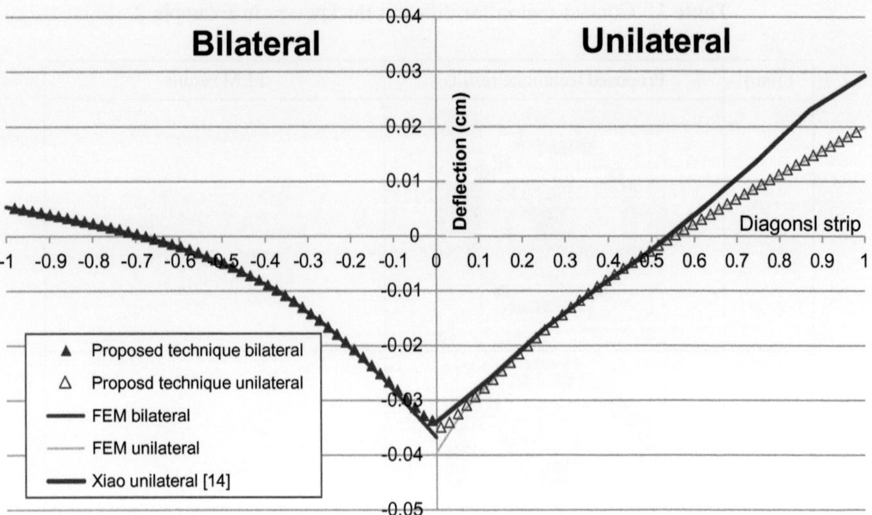

Fig. 9. Deflection along the diagonal strip for plate thickness = 30 cm for all models.

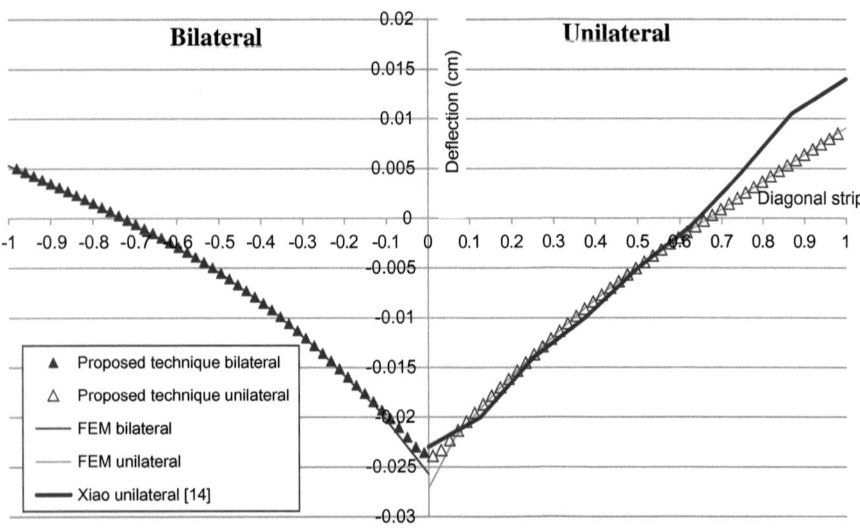

Fig. 10. Deflection along the diagonal strip for plate thickness = 40 cm for all models.

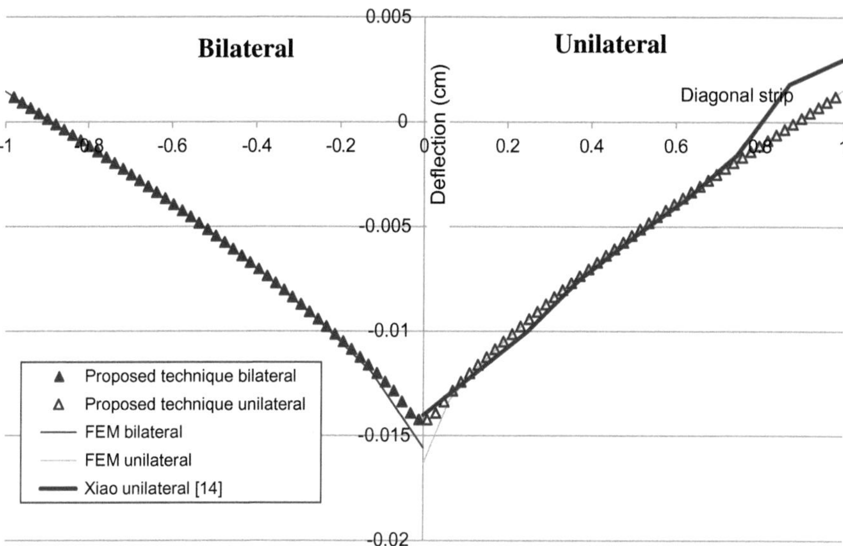

Fig. 11. Deflection along the diagonal strip for plate thickness = 60 cm for all models.

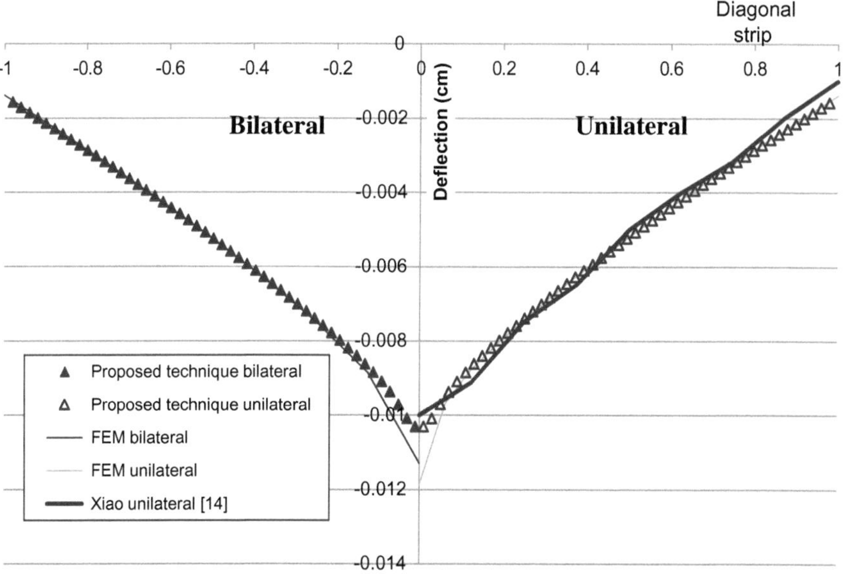

Fig. 12. Deflection along the diagonal strip for plate thickness = 80 cm for all models.

5 Conclusions

In this paper, solving plates on tensionless Winkler soil is presented using an efficient technique. The plate stiffness matrix is extracted using boundary element formulation. An iterative technique is used to eliminate the tensile stresses.

Several examples are presented and from results it can be concluded that the main advantages of this technique is:

1. The simplicity of dealing with the nonlinearity nature of the problem.
2. Avoiding the stresses concentration zones appears in finite element solutions by using the advantages of boundary element formulation.
3. Suitable to be extended to include solving plates on Winkler elastic-plastic foundations.
4. Accurate results near corners unlike in the formulation of Xiao [14] as shown in the examples of this paper.
5. Solving any geometry and any boundary conditions of the plate.

Acknowledgements. This project was supported financially by the Science and Technology Development Fund (STDF), Egypt, Grant No 14910. The authors would like to acknowledge the support of (STDF).

References

1. Katsikadelis, J.T., Armenakas, A.E.: Plates on elastic foundation by BIE method. J. Eng. Mech. **110**, 1086–1105 (1984)
2. Reissner, E.: On bending of elastic plates. Quart. Appl. Mathe. **5**, 55–68 (1947)
3. Vander Weeën, F.: Application of the boundary integral equation method to Reissner's plate model. Int. J. Numer. Methods Eng. **18**, 1–10 (1982)
4. Rashed, Y.F., Aliabadi, M.H., Brebbia, C.A.: The boundary element method for thick plates on a Winkler foundation. Int. J. Numer. Methods Eng. **41**, 1435–1462 (1998)
5. Rashed, Y.F., Aliabadi, M.H.: Boundary element analysis of buildings foundation plates. Eng. Anal. Boundary Elem. **24**, 201–206 (2000)
6. Rashed, Y.F.: A boundary/domain element method for analysis of building raft foundations. Eng. Anal. Boundary Elem. **29**, 859–877 (2005)
7. Weitsman, Y.: On foundations that react on compression only. ASME J. Appl. Mech. **37**, 1019–1030 (1970)
8. Celep, Z.: Rectangular plates resting on tensionless elastic foundation. J. Eng. Mech. **114**, 2083–2092 (1988)
9. Hui, L., Dempsey, J.P.: Unbonded contact of a square plate on an elastic half-space or a Winkler foundation. ASME J. Appl. Mech. **55**, 430–436 (1988)
10. Sapountzakis, E.J., Katsikadelis, J.T.: Unilaterally supported plates on elastic foundations by the boundary element method. ASME J. Appl. Mech. **59**, 580–586 (1992)
11. Kexin, M., Suying, F., Xiao, J.: A BEM solution for plates on elastic half-space with unilateral contact. Eng. Anal. Boundary Elem. **23**, 189–194 (1999)
12. Silva, A.R.D., Silveira, R.A.M., GonÇalves, P.B.: Numerical methods for analysis of plates on tensionless elastic foundations. Int. J. Solids Struct. **38**, 2083–2100 (2001)

13. Jirousek, J., Zieliński, A.P., Wróblewski, A.: T–element analysis of plates on unilateral elastic Winkler–type foundation. Comput. Assist. Mech. Eng. Sci. **8**, 343–358 (2001)
14. Xiao, J.R.: Boundary element analysis of unilateral supported Reissner plates on elastic foundations. Comput. Mech. **27**, 1–10 (2001)
15. Shen, H.S., Yu, L.: Nonlinear bending behavior of Reissner-Mindlin plates with free edges resting on tensionless elastic foundations. Int. J. Solids Struct. **41**, 4809–4825 (2004)
16. Silveira, R.A.M., Pereira, W.L.A., GonÇalves, P.B.: Nonlinear analysis of structural elements under unilateral contact constrains by a Ritz type approach. Int. J. Solids Struct. **45**, 2629–2650 (2008)
17. Torbacki, W.: Numerical analysis of beams on unilateral elastic foundation. Arch. Mater. Sci. Eng. **29**(2), 109–112 (2009)
18. Buczkowski, R., Torbacki, W.: Finite element analysis of plate on layered tensionless foundations. Arch. Civil Eng. LvI, 3 (2009)
19. Kongtong, P., Sukawat, D.: Coupled integral equations for uniformly loaded rectangular plates resting on unilateral supports. Int. J. Math. Anal. **7**, 847–862 (2013)

Efficient BEM Formulation for Analysis of Plates on Tensionless Half Space

Marina Reda[1] and Youssef F. Rashed[1,2(✉)]

[1] Department of Structural Engineering, Faculty of Engineering,
Cairo University, Giza, Egypt
[2] Supreme Council of Universities, Giza, Egypt
yrashed@scu.eg

Abstract. A new boundary element formulation (BEM) for the analysis of thick plates on unilateral elastic half space (EHS) is presented. Plate stiffness matrix is computed at degrees of freedom postulated at the interaction domain zone of the elastic half space. Hence this stiffness matrix is added to the corresponding stiffness matrix of the elastic half space. A condensation process is carried out in the elastic half space stiffness matrix to eliminate DOF with tensile stresses. Initially the plate is solved under the total applied load. An iterative process is performed to obtain the final contact zone. An example is presented to demonstrate the validity of the proposed technique and the results are compared to previously published results. The results of the present analysis are more proven to be more accurate than those obtained from previously published compared to finite element (FEM) results.

1 Introduction

The analysis of plates on nonlinear unilateral foundations is an important topic in structural engineering and has several applications in practical structural and geotechnical engineering. Analysis and research tackling this topic are widely spreading. The methods of analysis are developing to include the analysis of plates taking into account the underneath soil with its realistic physical presentation as half space, in addition to implementing the non-linear behavior. Since the soil is unable to carry any tensile stress, these stresses lead to a separation between the soil and the raft. Hence determining the contact region is a critical issue for the realistic modeling of the soil structure interaction problem.

Due to the importance of the problem, extensive research work had been carried out and published in the preceding decades. Moreover, different methods of analysis have been developed as the analytical and the numerical methods. The numerical methods had been highly attacked recently due to their higher potential solving complex problems as the Finite Element Method (FEM) and the Boundary Element Method (BEM). Concerning the FEM, Cheung and Zienkiewicz [1] obtained the FEM solution for the analysis of foundation structures on an elastic half space. Wardle and Fraser [2] extended the approach of Cheung and Zienkiewicz to include a multi-layered soil with isotropic and cross-anisotropic properties. Svec and Gladwell [3] developed an improved method for the analysis of a thin plastic plate on an elastic half space and

© Springer International Publishing AG 2018
M. Bouassida and M.A. Meguid (eds.), *Ground Improvement and Earth Structures*,
Sustainable Civil Infrastructures, DOI 10.1007/978-3-319-63889-8_12

Boussinesq's equation was used. Silva et al. [4] presented a FEM solution for elastic plates resting on Winkler and EHS foundations. Li and Dempsey [5] presented unbonded contact of a square plate on elastic Winkler or EHS foundation. Sharma et al. [6] used the finite element method for the analysis of rafts of any shape resting on an elastic half space. The vertical deflection at the node due to the contact pressure applied on the element was obtained by Boussinesq's solution. On the other hand, Sapountzakis and Katsikadelis [7] presented a boundary element solution for unilaterally supported plates resting on a homogenous or non-homogenous elastic Winkler foundation. Hu and Hartley [8] extended the approach of analysis of unilaterally supported plates to incorporate an elastic half space model which is based on the soil modulus of elasticity into the analysis. Xiao [9] presented a solution for thick rectangular plate with free edges on unilateral Winkler and EHS foundation. These two numerical methods are effectively powerful in the structural framework. However, it is to be mentioned that the FEM – when compared to the BEM – is a computationally time-demanding method of solution. In addition to the stress concentration that emerges at the location of the concentrated point loads applied on joints that form some sorts of singularities as will be demonstrated in the presented example.

Solving plates on tensionless foundation can be categorized into two main approaches. The first approach is based on solving a system of non-linear equations and via an algorithm coupled with an optimization technique. However, the second approach is based on the solution using an iterative process which will be acquainted with along this paper. The aim of this work is to establish a BEM formulation for the analysis of plates on unilateral (tensionless) half space foundations using the second approach iterative technique. The present paper relies on the general concept of the BEM theory necessary for the extraction of the plate stiffness matrix and its load vector. In addition, Mindlin's equation [10] is used for attaining the stiffness matrix of the EHS. Condensation method coupled with an iterative approach to attend the real contact zone. Proposed technique has the main advantage of the BEM in modeling the raft, in addition to solving system of equations adequate with iterative procedure. The presented example demonstrated the advantages of the proposed technique compared to previously published papers and FEM results.

2   Bem Formulation for Plate on Elastic Foundation

2.1   Plate Stiffness Matrix

The direct boundary integral equation of plate is presented according to Reissner plate bending theory [11]. Consider an arbitrary plate domain Ω and boundary Γ with internal supporting cells as shown in Fig. 1.

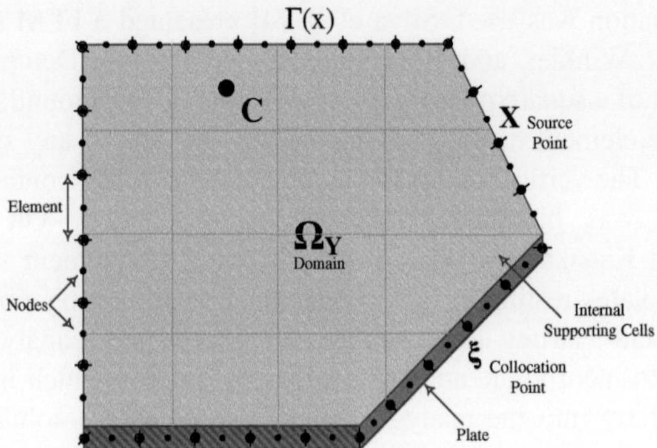

Fig. 1. Arbitrary plate for the BEM formulation.

The integral equation can be presented as follows:

$$C_{ij}(\xi)\,u_j(\xi) + \int_{\Gamma(x)} T_{ij}(\xi,x)\,u_j(x)d\Gamma(x) = \int_{\Gamma(x)} U_{ij}(\xi,x)\,t_j(x)d\Gamma(x)$$

$$+ \int_{\Gamma(x)} [V_{i,n}(\xi,x) - \frac{\nu}{(1-\nu)\lambda^2}U_{i\alpha}(\xi,x)]qd\Gamma(x) \tag{1}$$

Where $T_{ij}(\xi,x)$, $U_{ij}(\xi,x)$ are the two-point fundamental solution kernels for tractions and displacements respectively. The two points ξ and x are the source and the field points respectively. $u_j(x)$ and $t_j(x)$ denote the boundary generalized displacements and tractions. $C_{ij}(\xi)$ is the jump term and the kernel $V_i(\xi,x)$ is a suitable particular solution to represent domain loading. The symbols ν and λ denote the plate Poisson's ratio and shear factor. q denotes the intensity of domain loading.

Due to the presence of internal cells, the integral equation will be extended with the additive term of the internal cells.

$$C_{ij}(\xi)u_j(\xi) + \int_{\Gamma(x)} T_{ij}(\xi,x)\,u_j(x)d\Gamma(x) = \int_{\Gamma(x)} U_{ij}(\xi,x)t_j(x)d\Gamma(x)$$

$$+ \int_{\Gamma(x)} [V_{i,n}(\xi,x) - \frac{\nu}{(1-\nu)\lambda^2}U_{i\alpha}(\xi,x)]qd\Gamma(x) \tag{2}$$

$$+ \sum_{Nc} \int_{\Omega(y)} [U_{ik}(\xi,y) - \frac{\nu}{(1-\nu)\lambda^2}U_{i\alpha,\alpha}(\xi,y)\delta_{3k}F_k(y)d\Omega(y)$$

Nc denotes the number of internal supporting cells that having domain Ω_c, F_k denotes the supporting cell two bending moments ($F_1 = M_{xx}$, $F_2 = M_{yy}$) and column vertical force ($F_3 = F$). The field point y denotes the point of the internal supporting cell center.

When solving the previous equation, additional unknowns will emerge due to the presence of the internal supporting cells. In order to solve the equation with the number of unknowns greater than the number of equations, a collocation scheme is carried out at each internal supporting cell to attain the fact that the number of equations balances the number of unknowns. The collocation scheme is carried out as follows:

$$
\begin{aligned}
u_i(Y) + &\int_{\Gamma(x)} T_{ij}(Y,x)\,u_j(x)d\Gamma(x) = \int_{\Gamma(x)} U_{ij}(Y,x)t_j(x)d\Gamma(x) \\
&+ \int_{\Gamma(x)} [V_{i,n}(Y,x) - \frac{\nu}{(1-\nu)\lambda^2}U_{i\alpha}(Y,x)]qd\Gamma(x) \\
&+ \sum_{Nc}\int_{\Omega(y)} [U_{ik}(Y,y) - \frac{\nu}{(1-\nu)\lambda^2}U_{i\alpha,\alpha}(Y,y)\delta_{3k}F_k(y)d\Omega(y)
\end{aligned}
\tag{3}
$$

Where Y is a new source point located at each cell center.

Using Eqs. (2) and (3) the matrix conducted from the previous equation will be as follows:

$$
\begin{bmatrix} [A]_{3N\times3N} & [A_2]_{3N\times3Nc} & [0]_{3N\times3Nc} \\ [A_1]_{3Nc\times3N} & [A_3]_{3Nc\times3Nc} & [I]_{3Nc\times3Nc} \end{bmatrix}
\begin{Bmatrix} \{u/t\}_{3N\times1} \\ \{F_c\}_{3Nc\times1} \\ \{u_c\}_{3Nc\times1} \end{Bmatrix}
=
\begin{Bmatrix} \{RHS1\}_{3N\times1} \\ \{RHS2\}_{3Nc\times1} \end{Bmatrix}
\tag{4}
$$

Where $[A_1]_{3Nc\times3N}$, $[A_3]_{3Nc\times3Nc}$, and $\{RHS2\}_{3Nc\times1}$ contains the coefficients of the integrals presented in Eq. (3). $[0]_{3N\times3Nc}$, $[I]_{3Nc\times3Nc}$ are the null and identity matrices respectively. The vector $\{u_c\}_{3Nc\times1}$ contains the unknown values of internal supporting cell displacement.

The relation between $\{u_c\}_{3Nc\times1}$ and $\{F_c\}_{3Nc\times1}$ can be as follows:

$$
\{F_c\}_{3Nc\times1} = [K_c]_{3Nc\times3Nc}\{u_c\}_{3Nc\times1}
\tag{5}
$$

Where $[K_c]_{3Nc\times3Nc}$ is the stiffness matrix of internal supporting cells.

It is now of priority to get stiffness matrix of the plate from the previous matrix equation. This will be carried out by neglecting the domain load since the stiffness of the plate is fully independent on the load it carries. In addition to forcing the $\{u_c\}$ to be equals to one. The plate stiffness matrix as follows:

$$
\begin{bmatrix} [A]_{3N\times3N} & [A_2]_{3N\times3Nc} \\ [A_1]_{3Nc\times3N} & [A_3]_{3Nc\times3Nc} \end{bmatrix}
\begin{Bmatrix} \{u/t\}_{3N\times3Nc} \\ \{K_p\}_{3Nc\times3Nc} \end{Bmatrix}
=
\begin{Bmatrix} \{0\}_{3N\times3Nc} \\ -\{I\}_{3Nc\times3Nc} \end{Bmatrix}
\tag{6}
$$

Where $\{K_p\}_{3Ncx3Nc}$ is the required plate stiffness matrix.

2.2   Foundation Stiffness Matrix

The soil is modeled as elastic half space (EHS) [12]. EHS boundary surface is discretized into a number of cells N_{cb} corresponding to degrees of freedom postulated at the interaction domain zone with the plate as shown in Fig. 2.

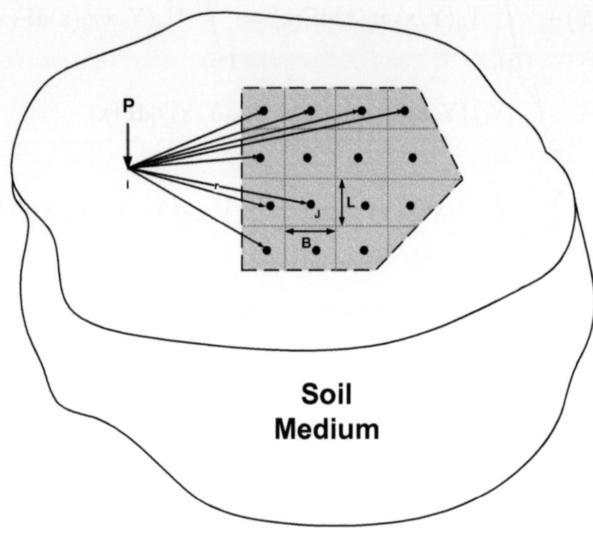

Fig. 2. Elastic half space presentation model.

The flexibility matrix of the EHS is computed at center of each cell using Mindlin's [10] equation.

$$w(i,i) = \frac{P}{8\pi G\,(1-v)B}\left[(3-4v)\left(\beta\ln\left(\frac{1+\sqrt{1+\beta^2}}{\beta}\right) + \ln\left(\beta+\sqrt{1+\beta^2}\right)\right)\right.$$
$$\left. + (5-12v-8v^2)\left(\beta\ln\left(\frac{1+\sqrt{1+\beta^2}}{\beta}\right) + \ln\left(\beta+\sqrt{1+\beta^2}\right)\right)\right]$$

$$(7)$$

Where $w_{(i,j)}$ is the deflection at point j due to load P at point i. v is the Poisson's ratio of the soil. G is the shear modulus of the soil and r is the distance between point i and j. B is the width of the cell and β is the ratio between width and length of the cell B/L. using Eq. 7 the flexibility matrix $[F_{Soil}]$ of the half space is calculated. Stiffness matrix of EHS is the inverse of computed flexibility matrix.

2.3   Coupling Between the Plate and the Foundation

This stiffness matrix is added to the corresponding stiffness matrix of the elastic half space. In order to couple between plate and foundation stiffness's, the rotational degrees of freedom of the plate are condensed to produce plate stiffness matrix $[K_{Plate}]_{Nc \times Nc}$. Assembly technique is used to couple between condensed plate stiffness matrix and underneath EHS stiffness matrix as follows:

$$[K_{Overall}]_{Nc \times Nc} = [K_{Plate}]_{Nc \times Nc} + [K_{Soil}]_{Nc \times Nc} \tag{8}$$

Where $[K_{Overall}]_{Nc \times Nc}$ is the overall stiffness of the plate rested on elastic foundation. $[K_{Soil}]_{Ncb \times Ncb}$ is the soil stiffness matrix.

In addition condensed load vector of plate is computed $\{P_{plate}\}_{Nc \times 1}$. The overall equilibrium equation can be written as follows:

$$[K_{Overall}]_{Nc \times Nc} * \{u_{plate}\}_{Nc \times 1} = \{P_{plate}\}_{Nc \times 1} \tag{9}$$

3   Proposed Technique

The present study carries this work forward by providing a technique of solving plates resting on EHS foundations using the iterative approach. Equation 9 is coupled with an iterative process of elimination of the tensile stresses emerging as shown in Fig. 3 until reaching the final contact zone (No tensile stresses emerge). A computer code is carried out to achieve the optimum solution.

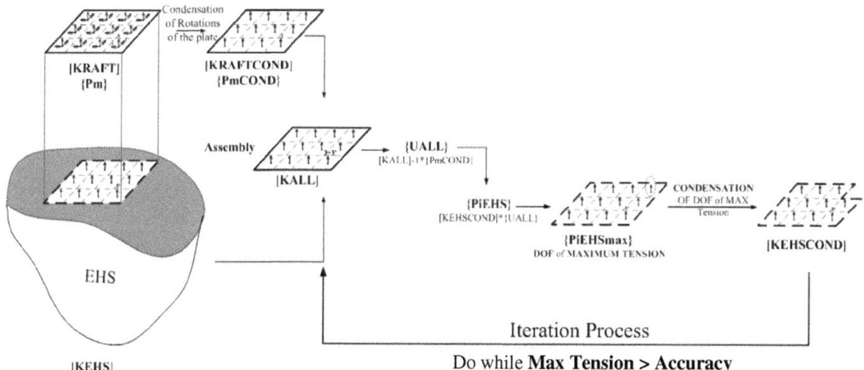

Fig. 3. Proposed technique process.

4 Examples

4.1 Free Edge Square Plate on Tensionless Half Space

Consider a square free edges plate resting on elastic half space. The plate is subjected to a concentrated load P = 1000 N at its center as shown in Fig. 4. The side of plate is a = 100 cm, the Young's modulus is E = 2.0×10^6 N/cm^2 and the Poisson's ratio of the plate material is v = 0.3. the Plate thickness is 2 cm. The soil modulus is E_s = 3000 N/cm^2 and its Poisson's ratio is v_s = 0.4. The same example was carried out modifying the soil modulus is E_s = 8600 N/cm^2 and different plate thicknesses (t) are employed 2, 4, 6, 8,10 cm. Figure 5 shows the BEM discretization model.

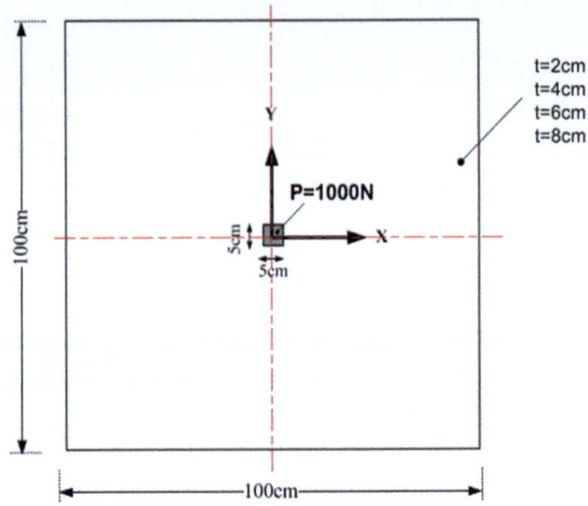

Fig. 4. Free edge square plate dimensions under patch load in example 1.

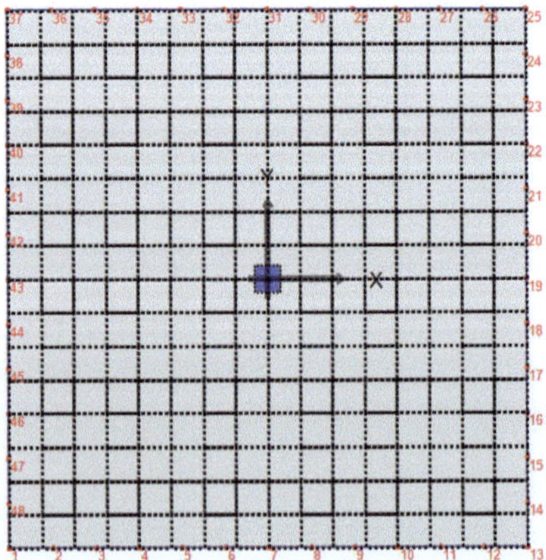

Fig. 5. BEM model discretization in example 1.

The results of the proposed technique are compared to the results obtained by Xiao [8] for both Young's modulus. Figures 7 and 8 demonstrate the deflection and soil pressure along diagonal strip (indicated in Fig. 6) respectively for the $E_s = 3000$ N/cm^2, bilateral and unilateral results are plotted also on these Figures for the sake of comparison. In order to not accurate results near corners in deflection results, FEM model is carried out. The deflection is in a good agreement with results of Xiao [9] except at corners, at which some inaccurate results due to singularity problems near boundary elements appear in the formulation of Xiao [9]. It can be seen that the

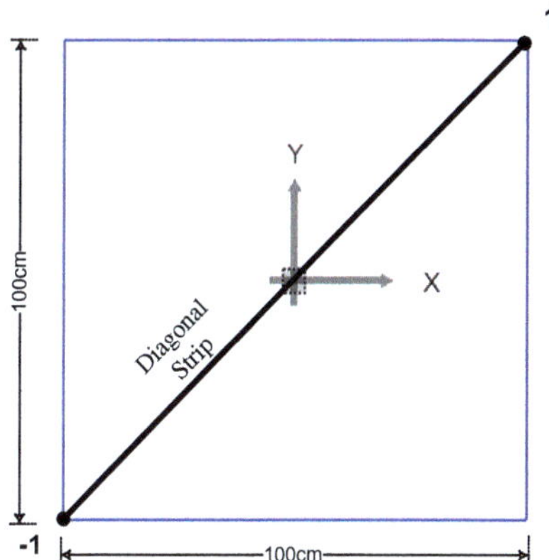

Fig. 6. Diagonal strip presentation for example 1.

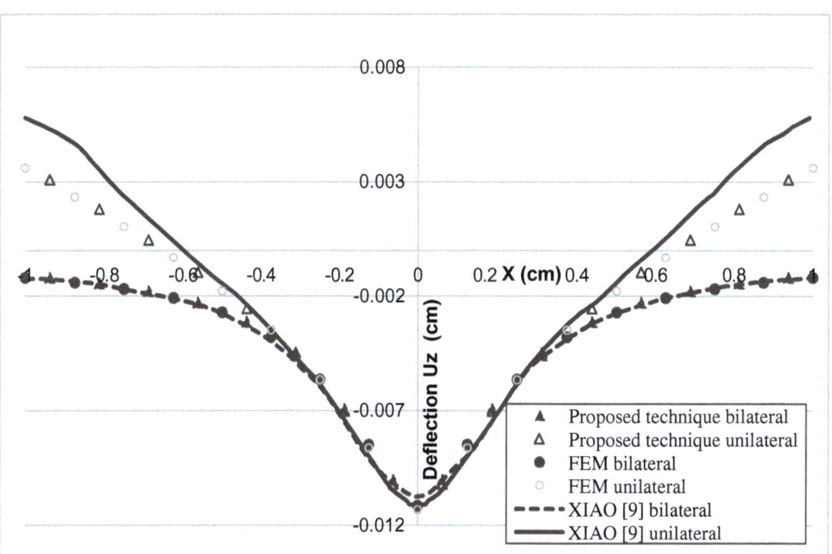

Fig. 7. Deflection along the diagonal strip for plate thickness = 2 cm.

presented formulation results are in good agreement with results from FEM. Figure 9 demonstrate comparison of deflection along the same diagonal strip of the proposed technique with Xiao [9] results for different plate thicknesses with $E_s = 8600$ N/cm^2. Figure 10 demonstrate the contact zones, for the different thicknesses, for both the proposed technique and Xiao [9].

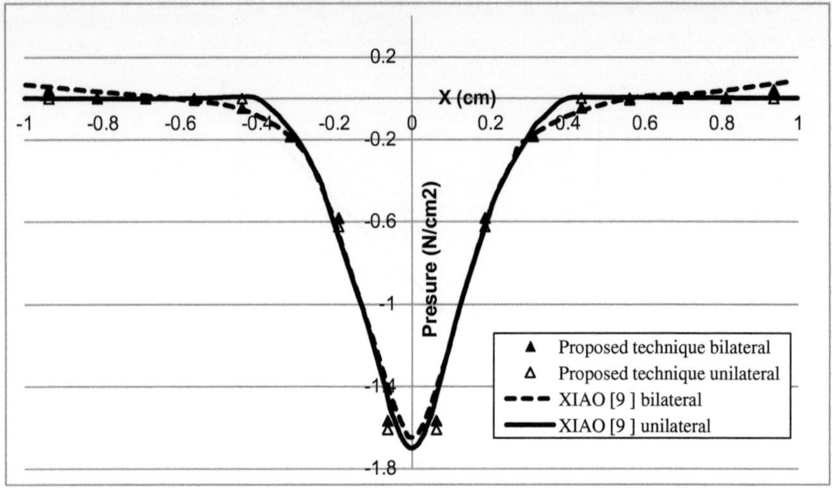

Fig. 8. Contact pressure along the diagonal strip for plate thickness = 2 cm.

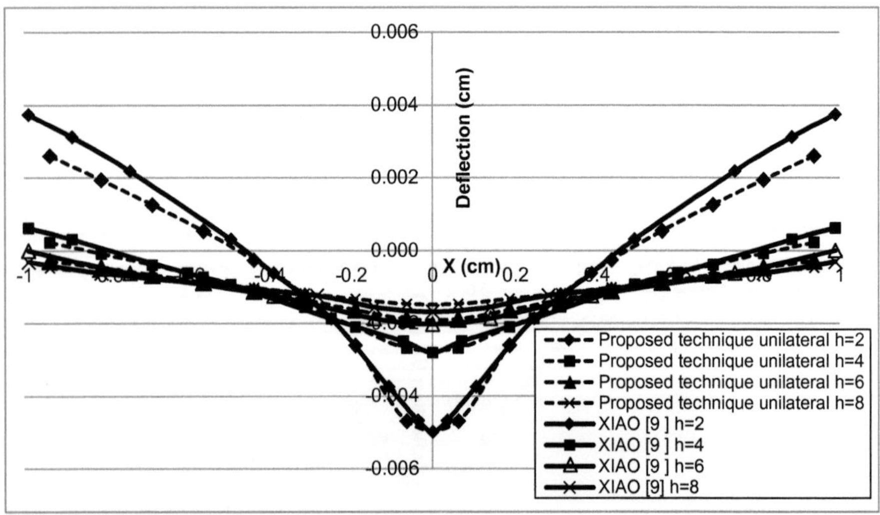

Fig. 9. Deflection along the diagonal strip for plate different thicknesses models (2,4,6,8 cm).

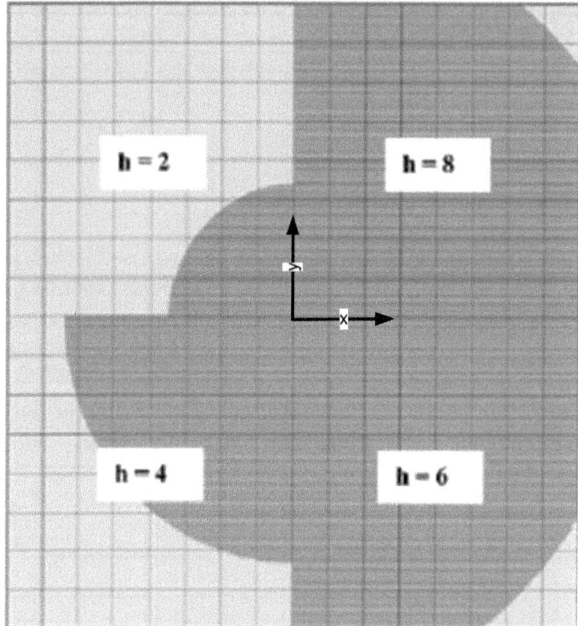

Xiao contact zone

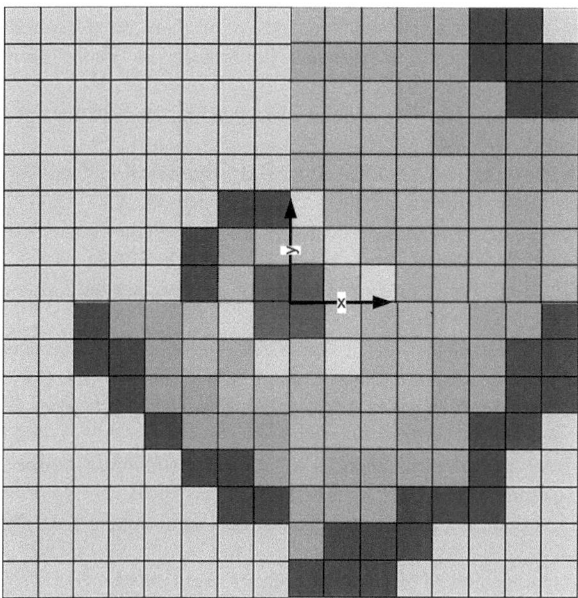

Proposed technique contact zone

Fig. 10. Unilateral – Contact zone for different models with thicknesses (2,4,6,8 cm).

It can be seen from results that proposed technique obtained the same contact area of Xiao [9]. This demonstrates the advantages of the proposed technique over the formulation of Xiao [9]. It is to be mentioned that the contact zone for the different thicknesses case was almost similar.

5 Conclusion

This precedent work presented the advantages of the proposed technique for analysis of plates resting on unilateral half space foundations. And the example presented showed the credibility of the proposed BEM coupled with the iterative condensation procedure. This formulation can deal with any geometry or any boundary condition of the plate.

Acknowledgements. This project was supported financially by the Science and Technology Development Fund (STDF), Egypt, Grant No 14910. The second author would like to acknowledge the support of (STDF).

References

1. Cheung, Y.K., Zienkiewicz, O.C.: Plates and tanks on elastic foundations-an application of finite element method. Int. J. Solids Struct. **1**, 451–461 (1965)
2. Wardle, L.J., et al.: Finite element analysis of a plate on a layered cross-anisotropic foundation. In: The First International Conference of Finite Element Methods in Engineering, pp. 565–578 (1974)
3. Svec, O.J., Gladwel, G.M.: A triangular plate bending element for contact problems. Int. J. Solids Struct. **9**, 435–446 (1973)
4. Silva, A.R.D., Silveira, R.A.M., Gonçalves, P.B.: Numerical methods for analysis of plates on tensionless elastic foundations. Int. J. Solids Struct. **38**, 2083–2100 (2001)
5. Li, H., Dempsey, J.P.: Unbonded contact of a square plate on an elastic half-space or a Winkler foundation. ASME J. Appl. Mech. **55**, 430–436 (1988)
6. Sharma, K.G., et al.: Finite element analysis of rafts resting on elastic half space. Indian Geotech. J. **14**(1), 28–39 (1984)
7. Sapountzakis, E.J., Katsikadelis, J.T.: Unilaterally supported plates on elastic foundations by the boundary element method. ASME J. Appl. Mech. **59**, 580–586 (1992)
8. Hu, C., Hartley, G.A.: Analysis of a thin plate on an elastic half-space. Comput. Struct. **52**(2), 227–235 (1994)
9. Xiao, J.R.: Boundary element analysis of unilateral supported reissner plates on elastic foundations. Comput. Mech. **27**, 1–10 (2001)
10. Mindlin, R.D.: Force at a point in the interior of a semi-infinite solid. Physics **7**, 195–202 (1936)
11. Reissner, E.: On bending of elastic plates. Quart. Appl. Mathe. **5**, 55–68 (1947)
12. Shaaban, A.M., Rashed, Y.F.: A coupled BEM-stiffness matrix approach for analysis of shear deformable plates on elastic half space. Eng. Anal. Boundary Elem. **37**(4), 699–707 (2013)

Numerical Analysis on the Performance of Fibre Reinforced Load Transfer Platform and Deep Mixing Columns Supported Embankments

Liet Chi Dang[1(✉)], Cong Chi Dang[2], and Hadi Khabbaz[1]

[1] School of Civil and Environmental Engineering,
University of Technology Sydney, Ultimo, NSW 2007, Australia
lietchi.dang@student.uts.edu.au,
hadi.khabbaz@uts.edu.au
[2] Department of Environment and Natural Resources,
Can Tho 910000, Hau Giang Province, Viet Nam
chicongctl912@gmail.com

Abstract. Deep cement mixing (DCM) columns are commonly employed as the most effective ground improvement approach in support of the road and railway embankments constructed over soft soils with low bearing capacity, insufficient shear strength and high compressibility. Finite element modeling is widely adopted to examine the performance of the road, railway and highway embankments during construction, post-construction as well as serviceability periods. Nevertheless, very limited studies have been conducted on the fibre reinforced load transfer platform (FRLTP) and DCM columns supported highway embankments constructed over soft clays. This paper presents a numerical investigation based on fine element method (FEM) to investigate the influence of fibre inclusion in the load transfer platform and DCM columns supported embankment on the stress transfer mechanism, overall and differential settlements, surface settlement versus horizontal distance from the centreline of embankment, settlement with depth, and variations of excess pore water pressure, which have been analysed and discussed in detail. The findings of this numerical analysis indicate that the FRLTP and DCM columns supported embankments can effectively alleviate the total settlement, excess pore water pressure and intensity of embankment load transfer to soft foundation soil, while considerably enhance the rigidity, stability and load transfer mechanism from the embankment to soil-cement columns.

1 Introduction

Many countries are experiencing a remarkable increase in development of essential infrastructure including roads and rail networks. Embankment construction is one of the most common methods used to provide better ground support for roads and railways. However, due to lack of readily available stiff foundation soils for such transport infrastructure projects, many embankments have to be founded on soft ground. This practice is highly risky because soft ground has low bearing capacity, insufficient shear

© Springer International Publishing AG 2018

M. Bouassida and M.A. Meguid (eds.), *Ground Improvement and Earth Structures*,
Sustainable Civil Infrastructures, DOI 10.1007/978-3-319-63889-8_13

strength and high compressibility. Therefore, in order to ensure the stability of embankment during construction process and long term services, appropriate ground improvement techniques are needed to be adopted in enhancing the engineering properties of soft soil or even in order to transfer embankment and traffic loads to a deeper and stiffer soil stratum.

A growing number of ground improvement approaches have been applied to improve soft soil properties to support embankment construction such as preloading with vertical drain application (Azari et al. 2016; Parsa-Pajouh et al. 2015; Liu and Rowe 2015), excavating the existing soft ground and substituting it with high shear strength and bearing capacity backfill soil, reducing embankment load by using lightweight fill material, constructing in stages and leaving time for consolidation, improving soft ground underneath embankment by chemical treatment (Dang et al. 2016a; Dang et al. 2015; Fatahi and Khabbaz 2013a), stone columns (Fatahi et al. 2012a) and geosynthetic-reinforced and pile supported earth platform (Dang et al. 2016b, c; Fatahi et al. 2012b; Fatahi et al. 2013; Han and Gabr 2002; Liu et al. 2007). On the top of those, the ground improvement technique using deep cement mixing columns has been widely used in construction practice because it provides an economical and fast ground improvement solution for the construction of road and highway embankment over soft soil (Chai et al. 2015; Jamsawang et al. 2016; Nguyen et al. 2014; Yapage et al. 2014, 2015; Fatahi and Khabbaz 2013b; Oliveira et al. 2011). In order to improve the soft soil characteristics such as bearing capacity, shear strength and compressibility, in wet mixing method, slurry cement is mixed with soft ground at a certainly designated water/cement ratio during the DCM columns construction process, while dry cement is mixed with in-situ soils to establish artificial DCM columns in dry mixing method.

Numerical modeling based on finite element method has been used as an effective tool in predicting the performance of DCM columns supported embankments founded on soft ground in terms of total and differential settlement, horizontal movement, rate of settlement, slope stability and degree of consolidation over a long period of time. The finite element numerical modeling can reasonably simulate the load transfer mechanism between DCM columns and foundation soil, the generation and dissipation of excess pore water pressure, consequently predict the total settlement, lateral displacement, bending moment of DCM columns with depth under the embankment using three-dimensional finite element model or even with an equivalent two-dimensional (2D) plane strain model (Chai et al. 2015; Tan et al. 2008; Yapage et al. 2014). Nevertheless, most of numerical investigations have been performed to investigate the behaviour of embankments over DCM columns (Chai et al. 2015; Jamsawang et al. 2016; Okyay and Dias 2010), the influence of georid reinforced embankments over piles/columns (Han and Gabr 2002; Liu et al. 2007; Yapage et al. 2014). Having not very much attention has been paid to the behaviour of load transfer platform (LTP) reinforced with lime-fibre-soil as a replacement of geosynthetics and DCM columns supported highway embankments constructed on soft ground.

This study presents a numerical analysis using a 2D finite element model with proper modified parameters to investigate the performance of DCM columns supported embankment with and without FRLTP placed over soft clays. Comparisons are undertaken to examine the effect of FRLTP on the embankment supported by DCM columns in terms of vertical deformation, variation of excess pore water pressure, total settlement with depth and horizontal distance from the centre of the embankment. The development of vertical effective stress and stress concentration ratio between DCM columns and foundation soil are also analysed and discussed in detail.

2   Case Study

A hypothetical construction of natural fibre-reinforced load transfer platform (FRLTP) and deep cement mixing (DCM) columns supported highway embankment over soft clay is considered for this numerical simulation. The embankment geometry is shown in Fig. 1 representing the only right half of the domain of the embankment since the embankment is symmetrical along its centreline. As can be seen in this figure, the embankment is 20 m wide and 4 m high with a 1 V:2H side slope. The embankment is made of sandy soil with a cohesion of 1 kPa, a friction angle of $30°$, and an average unit weight of 20 kN/m^3. It is constructed on a 3 m thick fill material overlaying an 11 m thick deposit of soft clay. This deposit soft clay overlies a 9 m thick medium stiff clay stratum followed by a 5 m thick stiff clay layer. The ground-water table is located at a depth of 1.5 m below the ground surface. Details of these soil layers are summarised in Table 1. A coir fibre-lime-soil layer having an effective cohesion and friction angle of 75 kPa and $42°$, respectively, an average unit weight of 12.5 kN/m^3, a Poisson's ratio of 0.32, a Young's modulus of 125.8 MPa, and tensile strength of 240 kPa as reported by Anggraini et al. (2015) is employed to reinforce load transfer platform of 0.5 m that is placed on the top of DCM columns.

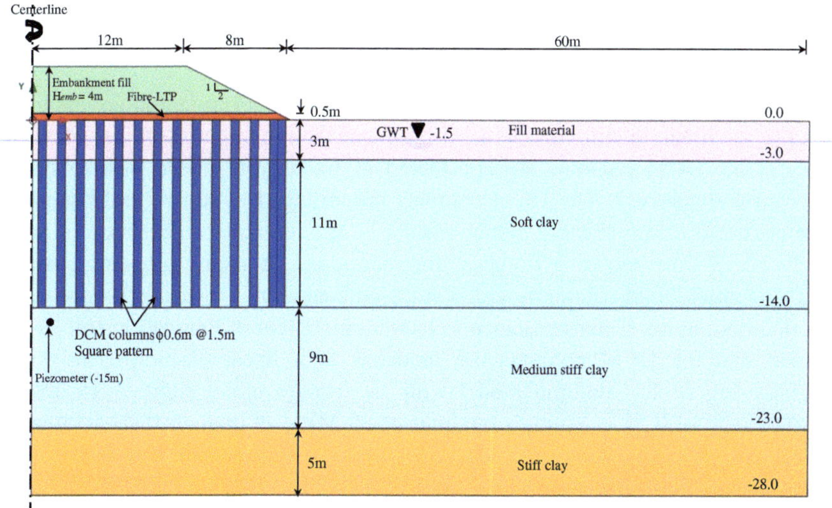

Fig. 1. Cross section of the fibre reinforced load transfer platform and DCM columns supported embankment

Table 1. Material properties of the embankment and subgrade soil layers

Parameters	Fill material	Soft clay	Medium stiff clay	Stiff clay	Fibre-lime-soil	Embankment fill	DCM columns
Depth (m)	0–3	3–14	14–23	23–28	–	–	–
Material model	MC[a]	SS[a]	SS	HS[a]	MC	MC	MC
Unit weight γ (kN/m^3)	20	14	16	20	12.5	20	15
Young's modulus, E (MPa)	30	–	–	–	125.8	20	80
Poisson's ratio, v	0.33	0.35	0.15	0.2	0.32	0.33	0.33
Effective cohesion, c' (kPa)	1	1	10	18	75	1	$c_u = 450$
Effective friction angle, φ' (°)	32	23	25	25	42	30	0
compression index, λ	–	0.18	0.12	–	–	–	–
Swelling index, κ	–	0.04	0.06	–	–	–	–
Secant stiffness, E_{50}^{ref} (MPa)	–	–	–	50	–	–	–
Tangential stiffness, E_{oed}^{ref} (MPa)	–	–	–	50	–	–	–
Unloading and reloading stiffness, E_{ur}^{ref} (MPa)	–	–	–	150	–	–	–
Power for stress-level dependency of stiffness, m	–	–	–	1	–	–	–
Over consolidation ratio, OCR	–	1.5	2	2.5	–	–	–
Permeability coefficient, k (m/day)	–	5×10^{-4}	2.5×10^{-4}	2.5×10^{-4}	–	–	5×10^{-4}
Material behaviour	Drained	Undrained	Undrained	Undrained	Undrained	Drained	Undrained type B

[a]MC: Mohr-Coulomb; SS: Soft Soil; HS: Hardening Soil

Deep cement mixing columns of 0.6 m diameter and 14 m length are assumed to extend to the medium stiff clay, which is stronger than the upper soft clay stratum. In this case, the DCM columns are considered as fixed type or end bearing columns supported embankment. The DCM columns are arranged in square grid pattern with 1.5 m centre to centre spacing, which results in an area replacement ratio of approximately 12.5% corresponding to these aforementioned geometric properties. An additional tangential DCM column is added at the embankment toe to increase the rigidity, while facilitating the better resistance to lateral movement of foundation soft soil. It is assumed that the DCM columns are modeled as a linear elastic-perfectly plastic material using Mohr-Coulomb model with an average unit weight of 15 kN/m^3, a Poisson's ratio of 0.33, a Young's modulus of 80 MPa, an undrained shear strength of $s_u = 450$ kPa, and tensile strength of 130 kPa.

The construction sequence of the embankment is assumed to be conducted in 8 lifts of 0.5 m increment including the placement of FRLTP. Each lift of embankment height increments is assumed to be completed in 30 days and followed by another 30 days waiting period of consolidation to dissipate the excess pore water pressure induced as a result of the increase in embankment height. At the completion of embankment

Table 2. Construction stages in the FEM simulation of construction procedure

Stage	Description	Thickness (m)	Duration (days)
1	Generation on the initial stresses (K_o- condition)	–	–
2	Installation of the DCM columns	–	–
3	Construction of a 0.5 m high embankment	0.5	10
3	Construction of a 1.0 m high embankment	0.5	45
4	Construction of a 1.5 m high embankment	0.5	50
5	Construction of a 2.0 m high embankment	0.5	30
6	Consolidation period of 30 days	–	30
7	Construction of a 2.5 m high embankment	0.5	30
8	Consolidation period of 30 days	–	30
9	Construction of a 3.0 m high embankment	0.5	30
10	Consolidation period of 30 days	–	30
11	Construction of a 3.5 m high embankment	0.5	30
12	Consolidation period of 30 days	–	30
13	Construction of a 4.0 m high embankment	0.5	30
14	Consolidation period of 2 years	–	730

construction, the consolidation is left for 2 years. Table 2 exhibits the simulated construction sequence of the highway embankment has been adopted in this numerical analysis.

3   Numerical Modeling

3.1   Finite Element Models and Parameters

A two-dimensional plane strain model was built using commercial finite element software PLAXIS 2D version 2015 adopting the equivalent 2D numerical analysis method proposed by previous researchers (Chai et al. 2015; Tan et al. 2008; Yapage et al. 2014) to simulate the performance of the natural fibre reinforced LTP and DCM columns supported highway embankment. The equivalent 2D model was selected because of less analysis time consuming, while generating results with reasonable accuracy. The DCM columns were simulated by continuous plane strain walls of 0.19 m thickness for the entire columns length of 14 m to maintain the same area of replacement ratio, taking into account of the equivalent either normal stiffness (EA) and bending rigidity (EI), while the centre to centre spacing between two adjacent walls in this numerical simulation was remained the same as the 1.5 m centre to centre spacing between two adjacent DCM columns.

With regard to the constitutive modeling, the DCM columns, as mentioned previously, were modeled as a linear elastic-perfectly plastic material using Mohr-Coulomb (MC) model (Huang and Han 2010; Liu and Rowe 2015; Jamsawang et al. 2016) with an average unit weight of 15 kN/m^3, a Poisson's ratio of 0.33, a Young's modulus of 80 MPa, an undrained shear strength of s_u = 450 kPa, and tensile strength of 130 kPa.

162   L.C. Dang et al.

Similarly, the fibre-lime reinforced load transfer platform, embankment and fill material were simulated using a linear elastic-perfectly plastic model with Mohr-Coulomb failure criterion. The Mohr-Coulomb material model requires Young's modulus (E), Poisson's ratio (v), effective cohesion (c'), angle of internal friction (φ'), and dilation angle (ψ). Subsequently, the soft soil and medium soft soil strata were represented by soft soil (SS) model. The required parameters for the SS model include compression index (λ), swelling index (κ), permeability coefficient (k), Young's modulus (E), Poisson's ratio (v), and shear strength properties (c',φ',ψ). It is assumed that the permeability coefficient of DCM columns was the same as surrounding soil. Finally, the hardening soil (HS) was adopted to model the stiff clay layer, which requires the input parameters consisting of secant stiffness $\left(E_{50}^{ref}\right)$ and unloading and reloading stiffness $\left(E_{ur}^{ref}\right)$ derived from conventional triaxial tests, tangential stiffness $\left(E_{oed}^{ref}\right)$ obtained from primary loading of one-dimensional consolidation tests, power for stress-level dependency of stiffness (m), permeability coefficient (k), and shear strength properties (c',φ', ψ). Assuming all subgrade foundation soils is isotropic, so the horizontal and vertical permeability of the subgrade soils are equal. A summary of the constitutive model parameters is presented in Table 1.

Referring to Fig. 2, only right half of the embankment is represented in this numerical simulation since the embankment is symmetrical along its centreline. The foundation soil was taken to 28 m depth from the ground surface overlying a stiff clay stratum. The horizontal length of the FEM model was taken to be 80 m, which was almost three times the half width of the embankment base, in order to minimize the boundary effect. All these boundaries were considered to be impermeable, and pore fluid flow was permitted only from the surface.

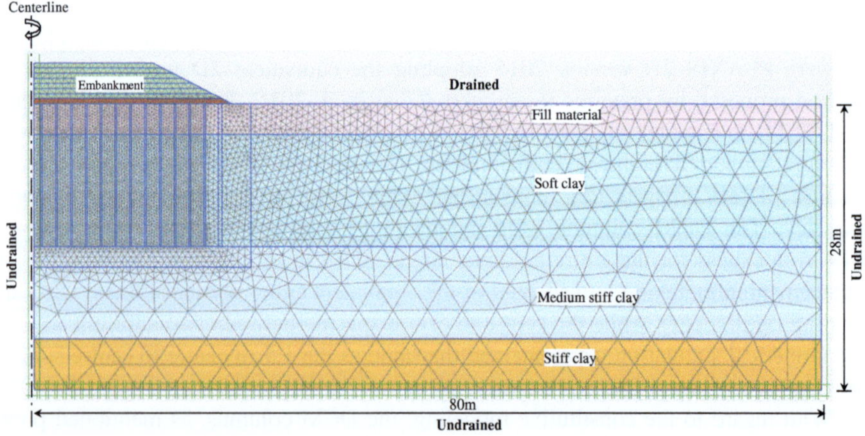

Fig. 2. Mesh and boundary conditions for a 2D FEM analysis of embankment

In this analysis, for the 2D plane strain FEM model, the horizontal displacement at the left and right boundaries was not permitted but vertical movement was allowed, whereas both the vertical and horizontal displacements were prevented at the bottom boundary. On one hand, due to the relatively high permeability, the embankment and fill material were assumed to behave in a drained condition. On the other hand, natural fibre reinforced load transfer platform, DCM columns, and other four foundation soils were assumed to act as undrained material. In this numerical analysis, fifteen-node triangular elements with excess pore water pressure degrees of freedom at all nodes were adopted to simulate the foundation soils, DCM columns and FRLTP, while fifteen nodes triangular elements without excess pore water pressure degrees of freedom at all nodes were applied to model the embankment and fill material.

3.2   Model Validation

A case study of DCM columns supported highway embankment reported by Jamsawang et al. (2016) was used to validate the proposed modeling approach employed in this numerical investigation. Jamsawang et al. (2016) reported the field measurement of settlement and excess pore water pressure as well as the predicted 3D modeling results for the investigated highway embankment. The detailed simulation procedure for the embankment analysis can be found in Jamsawang et al. (2016). In this numerical simulation, a comparison was undertaken between this current prediction results and those measured and predicted results of Jamsawang et al. (2016) on the base of the settlement at the centreline of the embankment base and the excess pore water pressure over time at a depth of 15 m from the embankment base, which was very closed to the DCM column tips. Figure 3(a) indicates the variation of settlement with time archived from this equivalent 2D finite element modeling is compared with the measured and predicted settlements reported by Jamsawang et al. (2016). It can be clearly observed that this proposed modeling agrees fairly well with those predicted and measured settlements of Jamsawang et al. (2016). Moreover, the excess pore water pressure with elapsed time obtained from this prediction as presented in Fig. 3(b) exhibits good agreement with the 3D modeling results of pore water pressure predicted by Jamsawang et al. (2016). Overall, the predicted results of this numerical analysis indicates that the equivalent 2D FEM model proposed in this study is suitable for simulating the DCM columns supported highway embankment over soft clay.

4   Analysis of Results and Discussion

4.1   Variation of Settlement and Excess Pore Water Pressure

Figure 4(a) and 4(b) display the development of the surface settlement between two adjacent DCM columns and the variation of excess pore water pressure at a depth of 15 m from the embankment base between two adjacent DCM columns supported embankment without and with fibre reinforced LTP, respectively. It can be seen from Fig. 4(a) that the surface settlement of the highway embankment over time increased significantly during the first period of 375 days due to the increase of embankment

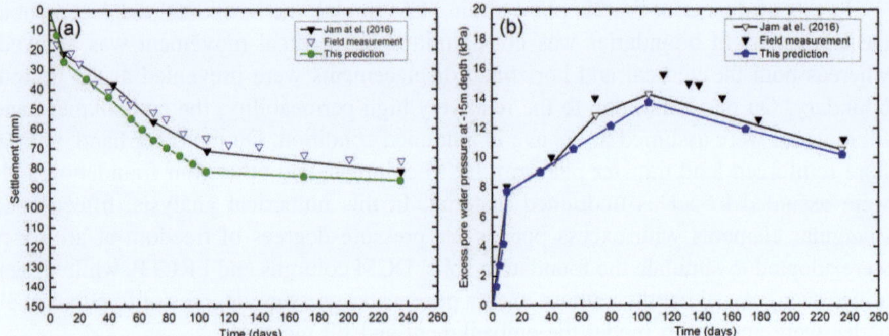

Fig. 3. Comparison between field measurement and 2D numerical prediction: (a) Settlement with time; (b) Excess pore water pressure with time

height and then followed by the gradual increase in the surface settlement up to two years after the completion of the embankments owing to the evolution of consolidation with elapsed time. However, the surface settlement at the centre of the DCM columns supported embankment with FRLTP was obviously smaller than that of the embankment without LTP, which can be clearly observed throughout the investigated duration. The maximum settlement at two years after the end of embankment construction for the FRLTP and DCM columns supported embankments was about 420 mm, which was relative smaller in compared with the 450 mm settlement of the DCM columns supported embankment without LTP. In addition, the smaller amount of excess pore water pressure generated by the embankment load versus time as presented in Fig. 4(b) is strongly supported for the smaller settlement of the DCM columns supported highway embankment with fibre reinforced LTP. The reduced settlement and excess pore water pressure on the FRLTP and DCM columns supported embankment could be attributed to the inclusion of FRLTP facilitating the effective improvement of load transfer mechanism between DCM columns and foundation soft soil as well as improving the arching effect with the increases in embankment height.

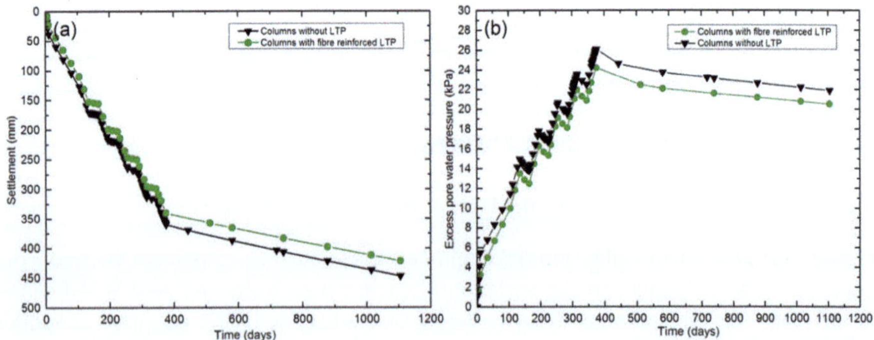

Fig. 4. Development of settlement and excess pore water pressure with time: (a) settlement; (b) excess pore water pressure

4.2   Load Transfer Mechanism Between Columns and Foundation Soil

Figure 5(a) and 5(b) reveal the development of the increase in embankment load induced vertical effective stress on the DCM column head at the centre of the highway embankment and stress concentration ratio (SCR) with time, respectively. It can be noted that the SCR is defined as the ratio of vertical effective stress on the DCM columns head to vertical effective stress applied on foundation soil between two adjacent DCM columns. By referring to SCR, the quantity of embankment load transfer to the DCM columns can be estimated. As can be seen from Fig. 5(a), the application of fibre reinforced LTP and DCM columns supported embankment resulted in the remarkable increase in the vertical effective stress induced by the embankment load on the DCM column head in comparison with that of the DCM columns supported embankment without LTP. The increase in the vertical effective stress on the DCM column head supported embankment with FRLTP was more noticeable with the higher increase in the embankment height and remained almost constant during the two year consolidation period after the completion of the embankment construction. The increased amount of vertical effective stress on columns head for the DCM column-embankment with FRLTP was approximately 25% greater than that of the DCM column-embankment without LTP. The similar trend can be observed in Fig. 5 (b) for the evolution of stress concentration ratio versus elapsed time. To be more specific, the DCM columns supported embankment with FRLTP yielded a considerably greater SCR value than the DCM column-embankment without LTP. For example, the SCR value of the DCM columns supported embankment with FRLTP was roughly 3.70 at the end of embankment construction compared to the SCR of 2.45 for the DCM column-embankment without LTP. The SCR improvement of the DCM columns supported embankment with FRLTP was considered as a direct consequence of the reduction of embankment load transfer to the subgrade foundation soft clay.

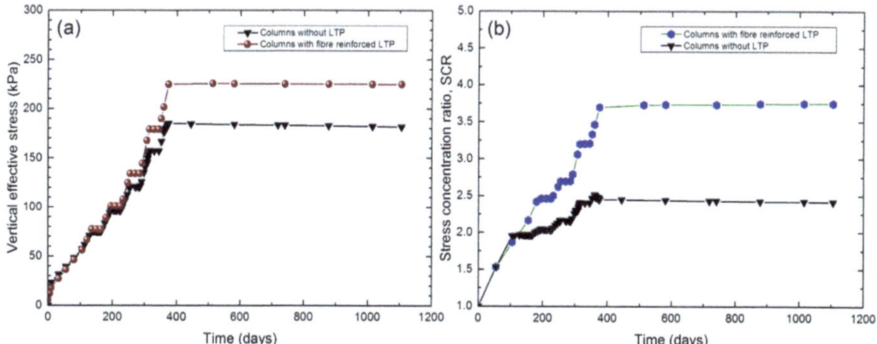

Fig. 5. Development of vertical effective stress on the top of DCM columns and stress concentration ratio with time: (a) vertical effective stress; (b) stress concentration ratio

4.3   Foundation Surface Settlement Distance from Centreline

The effect of the natural fibre reinforced LTP on the total settlement of foundation surface obtained at depth of 0.05 m with regard to horizontal distance from the centreline of the DCM columns supported embankment at the two years consolidation since the completion of embankment construction is illustrated in Fig. 6 for further comparison. It can be obviously seen that the larger settlement was observed for the DCM columns supported embankment without LTP in the range of embankment width, which demonstrates that the FRLTP and DCM column system can alleviate the total settlement of foundation surface effectively; whereas the total surface settlement of the DCM column-embankment without and with FRLTP was almost the same beyond the embankment base. Moreover, the upward displacement (heave) was also observed away from the embankment width. The maximum heave was obtained at approximately 26 m distance from the centreline of the embankment as a consequence of the uneven stress distribution of subgrade foundation soils.

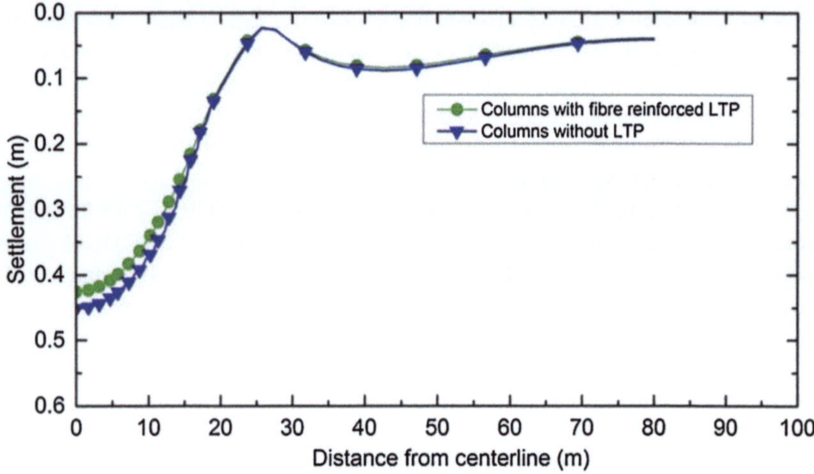

Fig. 6. Variation of surface settlement versus horizontal distance from centerline

4.4   Development of Settlement Versus Depth

Figure 7 exhibits the fluctuation of the total foundation settlement with different depth and embankment height under the centreline of the investigated embankment at two years post-construction. Generally, the total settlement of foundation soil surface increased between 200 mm and 450 mm with the increase in embankment height ranging from 2 m to 4 m, while linearly decreased with depth downward to a certain level of 23 m and then followed by the significant drop of the total settlement to almost zero ranging from 23 m to 28 m depth. However, the total settlement of foundation soil with depth was more pronounced and larger for the DCM columns supported embankment without LTP, observed in any given embankment heights, which indicates the effectiveness of the FRLTP and DCM columns supported embankment in improving the total settlement of foundation soil underneath the investigated embankment. The better improvement in the vertical deformation of the foundation soils could be due to the inclusion of FRLTP into the DCM column-embankment system, which enhanced the stiffness and arching effect of the entire embankment that facilitated the embankment load transfer to stronger, stiffer clay below.

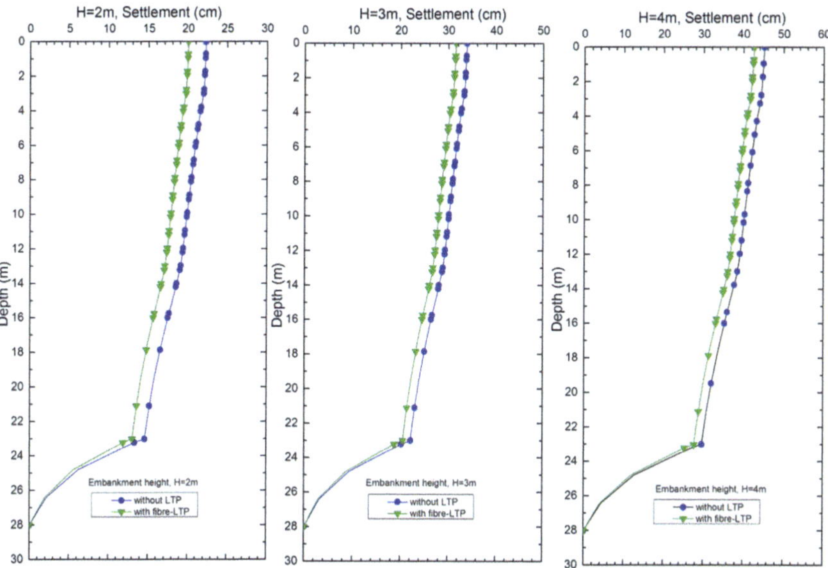

Fig. 7. Development of settlement with depth and embankment heights, H = 2, 3 & 4 m

5   Conclusions

In this numerical investigation, the simulation of fibre reinforced load transfer platform and deep cement mixing columns supported embankment was conducted using an equivalent 2D plane strain FEM model to examine the behaviour of the embankment without and with fibre reinforced load transfer platform (FRLTP). Generally, the application of natural fibre-lime reinforced load transfer platform assists the higher embankment load transfer to DCM columns, while minimises the embankment load transfer to foundation soil between two adjacent DCM columns, consequently miti-gates the generation of excess pore water pressure in the soft soil stratum. As a result, the utilisation of fibre reinforced LTP and DCM columns supported embankment prevents the total and differential settlements of the embankment effectively in com-parison with the DCM column-embankment without LTP. More importantly, this numerical investigation explores an interesting potential for making use of agricultural and industrial waste by-products such as coconut coir fibre, bagasse fibre and jute fibre as construction fill materials for sustainable development in the area of infrastructure foundations.

References

Azari, B., Fatahi, B., Khabbaz, H.: Assessment of the elastic-viscoplastic behavior of soft soils improved with vertical drains capturing reduced shear strength of a disturbed zone. Int. J. Geomech. **16**(1), B4014001 (2016)

Anggraini, V., Asadi, A., Huat, B.B.K., Nahazanan, H.: Performance of chemically treated natural fibres and lime in soft soil for the utilisation as pile-supported earth platform. Int. J. Geosynth. Ground Eng. **1**(3), 1–14 (2015)

Chai, J., Shrestha, S., Hino, T., Ding, W., Kamo, Y., Carter, J.: 2D and 3D analyses of an embankment on clay improved by soil–cement columns. Comput. Geotech. **68**, 28–37 (2015)

Dang, L., Hasan, H., Fatahi, B., Jones, R., Khabbaz, H.: Enhancing the engineering properties of expansive soil using bagasse ash and hydrated lime. Int. J. Geomate **11**(25), 2447–2454 (2016a)

Dang, L., Hasan, H., Fatahi, B., Khabbaz, H.: Influence of bagasse ash and hydrated lime on strength and mechanical behaviour of stabilised expansive soil. In: Côté, J., Allard, M. (eds.) GEOQuébec 2015, pp. 1–8. Québec City, Canada (2015)

Dang, L., Dang, C., Fatahi, B., Khabbaz, H.: Numerical assessment of fibre inclusion in a load transfer platform for pile-supported embankments over soft soil. In: Chen, D., Lee, J., Steyn, W.J. (eds.) Geo-China 2016, vol. GSP 266, pp. 148–55. ASCE (2016b)

Dang, L.C., Fatahi, B., Khabbaz, H.: Behaviour of expansive soils stabilized with hydrated lime and bagasse fibres. Procedia Eng. **143**, 658–65 (2016c)

Fatahi, B., Basack, S., Premananda, S., Khabbaz, H.: Settlement prediction and back analysis of Young's modulus and dilation angle of stone columns. Aust. J. Civil Eng. **10**(1), 67 (2012)

Fatahi, B., Khabbaz, H., Fatahi, B.: Mechanical characteristics of soft clay treated with fibre and cement. Geosynth. Int. **19**, 252–262 (2012)

Fatahi, B., Fatahi, B., Le, T.M., Khabbaz, H.: Small-strain properties of soft clay treated with fibre and cement. Geosynth. Int. **20**, 286–300 (2013)

Fatahi, B., Khabbaz, H.: Influence of fly ash and quicklime addition on behaviour of municipal solid wastes. J. Soils Sedim. **13**(7), 1201–1212 (2013a)

Fatahi, B., Khabbaz, H.: Predicting settlement of chemically stabilised landfills. Int. J. Geomate **5**(2), 700–705 (2013b)

Han, J., Gabr, M.: Numerical analysis of geosynthetic-reinforced and pile-supported earth platforms over soft soil. J. Geotech. Geoenviron. Eng. **128**(1), 44–53 (2002)

Huang, J., Han, J.: Two-dimensional parametric study of geosynthetic-reinforced column-supported embankments by coupled hydraulic and mechanical modeling. Comput. Geotech. **37**(5), 638–648 (2010)

Jamsawang, P., Yoobanpot, N., Thanasisathit, N., Voottipruex, P., Jongpradist, P.: Three-dimensional numerical analysis of a DCM column-supported highway embankment. Comput. Geotech. **72**, 42–56 (2016)

Liu, H., Ng, C., Fei, K.: Performance of a geogrid-reinforced and pile-supported highway embankment over soft clay: case study. J. Geotech. Geoenviron. Eng. **133**(12), 1483–1493 (2007)

Liu, K.W., Rowe, R.K.: Numerical modelling of prefabricated vertical drains and surcharge on reinforced floating column-supported embankment behaviour. Geotext. Geomembr. **43**(6), 493–505 (2015)

Nguyen, L.D., Fatahi, B., Khabbaz, H.: A constitutive model for cemented clays capturing cementation degradation. Int. J. Plast **56**, 1–18 (2014)

Okyay, U.S., Dias, D.: Use of lime and cement treated soils as pile supported load transfer platform. Eng. Geol. **114**(1–2), 34–44 (2010)

Venda Oliveira, P.J., Pinheiro, J.L.P., Correia, A.A.S.: Numerical analysis of an embankment built on soft soil reinforced with deep mixing columns: parametric study. Comput. Geotech. **38**(4), 566–576 (2011)

Parsa-Pajouh, A., Fatahi, B., Khabbaz, H.: Experimental and numerical investigations to evaluate two-dimensional modeling of vertical drain-assisted preloading. Int. J. Geomech. **16**(1), B4015003 (2015)

Tan, S., Tjahyono, S., Oo, K.: simplified plane-strain modeling of stone-column reinforced ground. J. Geotech. Geoenviron. Eng. **134**(2), 185–194 (2008)

Yapage, N., Liyanapathirana, D., Kelly, R., Poulos, H., Leo, C.: numerical modeling of an embankment over soft ground improved with deep cement mixed columns: case history. J. Geotech. Geoenviron. Eng. **140**(11), 04014062 (2014)

Yapage, N., Liyanapathirana, D., Poulos, H., Kelly, R., Leo, C.: Numerical modeling of geotextile-reinforced embankments over deep cement mixed columns incorporating strain-softening behavior of columns. Int. J. Geomech. **15**(2), 04014047 (2015)

Author Index

A
Abdel-Rahman, Khalid, 89
AbdelSalam, Sherif S., 46
Achmus, Martin, 89
Alexiew, Dimiter, 54
Almohd, Izzaldin M., 54
Azzam, Salem A., 46

B
Baba, Khadija, 66
Bahar, Ramdane, 121
Bahi, Lahcen, 66
Benazzoug, Mouloud, 121
Bhuse, Ankush, 20
Bouassida, Mounir, 100

D
Dang, Cong Chi, 157
Dang, Liet Chi, 157

E
El-Emam, Magdi, 34
El-Sherbiny, Rami M., 54

F
Farid, Ahmed Fady, 133
Frikha, Wissem, 100, 108

G
Ganesh, R., 1
Gerlach, Tim, 89

J
Jebali, Halima, 100
Jellali, Belgacem, 108

K
Khabbaz, Hadi, 157

M
Mane, A.S., 20

N
Nehab, Noura, 66

O
Ouadif, Latifa, 66

R
Rashed, Youssef F., 133, 146
Reda, Marina, 146

S
Sahoo, Jagdish Prasad, 1
Shete, Shubham, 20
Shokry, Beshoy M., 46

T
Touqan, Majid, 34

© Springer International Publishing AG 2018
M. Bouassida and M.A. Meguid (eds.), *Ground Improvement and Earth Structures*,
Sustainable Civil Infrastructures, DOI 10.1007/978-3-319-63889-8

MIX
Papier aus verantwortungsvollen Quellen
Paper from responsible sources
FSC® C105338

If you have any concerns about our products,
you can contact us on
ProductSafety@springernature.com

In case Publisher is established outside the EU,
the EU authorized representative is:
**Springer Nature Customer Service Center GmbH
Europaplatz 3, 69115 Heidelberg, Germany**

Printed by Libri Plureos GmbH
in Hamburg, Germany